Arsenic in Drinking Water

2001 UPDATE

Subcommittee to Update the 1999 Arsenic in Drinking Water Report

Committee on Toxicology

Board on Environmental Studies and Toxicology

Division on Earth and Life Studies

National Research Council

NATIONAL ACADEMY PRESS
Washington, D.C.

NATIONAL ACADEMY PRESS   2101 Constitution Ave., N.W.   Washington, D.C. 20418

NOTICE: The project that is the subject of this report was approved by the Governing Board of the National Research Council, whose members are drawn from the councils of the National Academy of Sciences, the National Academy of Engineering, and the Institute of Medicine. The members of the committee responsible for the report were chosen for their special competences and with regard for appropriate balance.

This project was supported by Cooperative Agreement No. X-82899401-0 between the National Academy of Sciences and the U.S. Environmental Protection Agency. Any opinions, findings, conclusions, or recommendations expressed in this publication are those of the author(s) and do not necessarily reflect the view of the organizations or agencies that provided support for this project.

Library of Congress Catalog Card Number: 2001097317
International Standard Book Number: 0-309-07629-3

Additional copies of this report are available from:

National Academy Press
2101 Constitution Ave., NW
Box 285
Washington, DC 20055

800-624-6242
202-334-3313 (in the Washington metropolitan area)
http://www.nap.edu

THE NATIONAL ACADEMIES
National Academy of Sciences
National Academy of Engineering
Institute of Medicine
National Research Council

The **National Academy of Sciences** is a private, nonprofit, self-perpetuating society of distinguished scholars engaged in scientific and engineering research, dedicated to the furtherance of science and technology and to their use for the general welfare. Upon the authority of the charter granted to it by the Congress in 1863, the Academy has a mandate that requires it to advise the federal government on scientific and technical matters. Dr. Bruce M. Alberts is president of the National Academy of Sciences.

The **National Academy of Engineering** was established in 1964, under the charter of the National Academy of Sciences, as a parallel organization of outstanding engineers. It is autonomous in its administration and in the selection of its members, sharing with the National Academy of Sciences the responsibility for advising the federal government. The National Academy of Engineering also sponsors engineering programs aimed at meeting national needs, encourages education and research, and recognizes the superior achievements of engineers. Dr. Wm. A. Wulf is president of the National Academy of Engineering.

The **Institute of Medicine** was established in 1970 by the National Academy of Sciences to secure the services of eminent members of appropriate professions in the examination of policy matters pertaining to the health of the public. The Institute acts under the responsibility given to the National Academy of Sciences by its congressional charter to be an adviser to the federal government and, upon its own initiative, to identify issues of medical care, research, and education. Dr. Kenneth I. Shine is president of the Institute of Medicine.

The **National Research Council** was organized by the National Academy of Sciences in 1916 to associate the broad community of science and technology with the Academy's purposes of furthering knowledge and advising the federal government. Functioning in accordance with general policies determined by the Academy, the Council has become the principal operating agency of both the National Academy of Sciences and the National Academy of Engineering in providing services to the government, the public, and the scientific and engineering communities. The Council is administered jointly by both Academies and the Institute of Medicine. Dr. Bruce M. Alberts and Dr. Wm. A. Wulf are chairman and vice chairman, respectively, of the National Research Council.

vii

Preface

Under the Safe Drinking Water Act of 1976, the U.S. Environmental Protection Agency (EPA) adopted an interim maximum contaminant level (MCL) for arsenic in drinking water of 50 micrograms per liter (μg/L) as part of the National Interim Primary Drinking Water Standards; that standard was originally set in 1942. EPA conducted a risk assessment for arsenic in drinking water in 1988. In 1996, the National Research Council (NRC) was asked to review independently the scientific database and evaluate the validity of that risk assessment. The NRC's 1999 report *Arsenic in Drinking Water,* concluded that "the current EPA MCL for arsenic in drinking water of 50 μg/L does not achieve EPA's goal for public health protection and therefore requires downward revision as promptly as possible." On January 22, 2001, EPA issued a pending standard MCL of 10 μg/L. Then on March 23, 2001, EPA delayed the effective date of the arsenic rule until further studies were conducted. In April, EPA's Office of Water requested that the NRC update its 1999 *Arsenic in Drinking Water* report.

In this report, the NRC's Subcommittee to Update the 1999 Arsenic in Drinking Water Report reviews the available toxicological, epidemiological, and risk assessment literature that has been published since the 1999 report. The subcommittee reviewed data for dose-response assessment and risk estimation; assessed whether the most recent EPA analysis is adequate for estimating an effective dose for a 1% response; determined whether EPA's analysis appropriately considers and characterizes the available data on the mode of action of arsenic and the information on dose-response and uncertainties when assessing the public-health impacts; and determined whether EPA's risk estimates for 3, 5, 10, and 20 µg/L of arsenic are consistent with available scientific information, including information from new studies.

This report has been reviewed in draft form by individuals chosen for their diverse perspectives and technical expertise, in accordance with procedures approved by the NRC's Report Review Committee. The purpose of this independent review is to provide candid and critical comments that will assist the institution in making its published report as sound as possible and to ensure that the report meets institutional standards for objectivity, evidence, and responsiveness to the study charge. The review comments and draft manuscript remain confidential to protect the integrity of the deliberative process. We wish to thank the following individuals for their review of this report: Bruce N. Ames, University of California, Berkeley; H. Vasken Aposhian, University of Arizona; Andrew A. Benson, University of California, San Diego; Kenneth G. Brown, Kenneth G. Brown, Inc.; Yvonne Dragan, Ohio State University; Marsha Ford, Carolinas Medical Center; Joshua W. Hamilton, Dartmouth College; Bruce P. Lanphear, University of Cincinnati; Denise Lewis, U.S. Department of Agriculture; Roger O. McClellan, Chemical Industry Institute of Toxicology; Mitchell Small, Carnegie Mellon University; Allan H. Smith, University of California, Berkeley; and Ronald Wyzga, Electric Power Research Institute. Although the reviewers listed above have provided many constructive comments and suggestions, they were not asked to endorse the conclusions or recommendations nor did they see the final draft of the report before its release. The review of this report was overseen by Gilbert S. Omenn, University of Michigan, and Frank H. Stillinger, Princeton University. Appointed by the NRC, they were responsible for making certain that an independent examination of this report was carried out in accordance with institutional procedures and that all review comments were carefully considered. Responsibility for the final content of this report rests entirely with the authoring committee and the institution.

The subcommittee gratefully acknowledges the following individuals for providing background information and for making presentations to the subcommittee: Christine Todd Whitman, Ephraim King, Diane Regas, James Taft, and Jeanette Wiltse, EPA; Richard Bull, EPA's Science Advisory Board Drinking Water Committee; Lawrence Bazel, Western Coalition of Arid States; Erik Olson, Natural Resources Defense Council; Allan Smith, University of California, Berkeley; Gerhard Stohrer, Risk Policy Center; Janice Yager, Electric Power Research Institute; Gunther Craun, City of Albuquerque; Kevin Bromberg, U.S. Small Business Administration; Bob Fensterheim,

Environmental Arsenic Council; Angela Logomasini, Competitive Enterprise Institute; Steven Lamm, Consultants in Epidemiology and Occupational Health, Inc.; Richard Nelson, Nebraska Department of Health and Human Services; and Tom Richichi, American Wood Preservers Institute. In addition, the subcommittee wants to give special thanks to individuals who provided data, further analysis, statistical analysis programs, and background information to the subcommittee: John Bennett, Andrew Schulman, Irene Dooley, and Charles Abernathy, EPA; Knashawn Morales, Harvard School of Public Health; Ming-Hui Chen, Worcester Polytechnic Institute; Claudia Hopenhayn-Rich, University of Kentucky; Michael Bates, University of California, Berkeley; and John Potter, Fred Hutchison Cancer Research Center.

The subcommittee is grateful for the assistance of the NRC staff for preparing the report. Staff members who contributed to this effort are Michelle Catlin, project director; James J. Reisa, director of the Board on Environmental Studies and Toxicology; Roberta M. Wedge, program director for risk analysis; Kulbir Bakshi, program director for the Committee on Toxicology; Ruth E. Crossgrove, managing editor; Mirsada Karalic-Loncarevic, information specialist; and Bryan P. Shipley, project assistant. Other staff members who contributed to this effort are Ellen Mantus, Abigail Stack, and Susan Martel, program officers; Leah Probst, Lucy Fusco, and Jessica Parker, project assistants; and Kelly Clark, editorial assistant.

Finally, I would especially like to thank all the members of the subcommittee for their exceptional efforts throughout the development of this report. The subcommittee members have gone above and beyond the call of duty to complete this report within a very short time frame.

Robert A. Goyer, M.D.
Chair, Subcommittee to
Update the 1999 Arsenic Report

Contents

Arsenic in Drinking Water:
2001 Update

Summary

The U.S. Environmental Protection Agency (EPA) is required under the Safe Drinking Water Act (SDWA) to establish the concentrations of contaminants permitted in public drinking-water supplies. The SDWA requires EPA to set two specific concentrations for each designated contaminant in drinking water—the maximum contaminant level goal (MCLG) and the maximum contaminant level (MCL). The MCLG is a health goal to be based on the best available, peer-reviewed scientific data. It is to be set at a concentration at which no known or anticipated adverse health effects occur, allowing for adequate margins of safety. The MCLG is not a regulatory requirement and might not be attainable with current technology or analytical methods. In contrast, the MCL is an enforceable standard that is required to be set as close to the MCLG as is technologically feasible, taking cost into consideration.

Following the 1976 enactment of the SDWA, EPA proposed, as part of the National Interim Primary Drinking Water Standards, an interim MCL of 50 micrograms per liter (μg/L) for arsenic. The U.S. Public Health Service originally set the 50-μg/L standard in 1942. In 1988, EPA conducted a risk assessment for arsenic in drinking water and, in 1996, requested that the National Research Council (NRC), the operating arm of the National Academy of Sciences and the National Academy of Engineering, independently review the scientific database and evaluate the scientific validity of that risk assessment. In response to that request, the NRC published *Arsenic in Drinking Water* in 1999. Following that report, EPA proposed an arsenic standard

1

of 5 µg/L in the *Federal Register*. After review by EPA's Science Advisory Board (SAB) and a period of public comment, EPA issued a pending standard of 10 µg/L on January 22, 2001. That pending standard was primarily based on dose-response models and extrapolation from a cancer study of a Taiwanese population exposed to high concentrations of arsenic in its drinking water. On March 23, 2001, EPA published a notice that delayed the effective date of the arsenic rule pending further study of options for revising the MCL for arsenic. To incorporate the most recent scientific research into its decision, EPA's Office of Water subsequently requested that the NRC independently review studies on the health effects of arsenic published since the 1999 NRC report.

CHARGE TO THE SUBCOMMITTEE

In response to EPA's request, the NRC assigned the project to the Committee on Toxicology (COT) and convened the Subcommittee to Update the 1999 *Arsenic in Drinking Water* Report. The members selected by the NRC to serve on this subcommittee have expertise in epidemiology, cellular and molecular toxicology, biostatistics and modeling, risk assessment, uncertainty analyses, and public health. Five of the nine members of the subcommittee also served on the earlier NRC Subcommittee on Arsenic in Drinking Water. The 2001 subcommittee was charged with the task of preparing a report updating the scientific analyses, uncertainties, and findings of the 1999 report on the basis of relevant toxicological and health-effects studies published and relevant data developed since the 1999 NRC report and to evaluate the analyses subsequently conducted by EPA in support of its regulatory decision-making for arsenic in drinking water. The subcommittee was charged and constituted to address only scientific topics relevant to toxicological risk and health effects of arsenic. It was not asked to address questions of economics, cost-benefit assessment, control technology, exposure assessment in U.S. populations, or regulatory decision-making. The subcommittee performed the following tasks in response to its charge:

- Determine whether data from the 1988, 1989, and 1992 Taiwanese studies remain the best data for dose-response assessment and risk estimation.
- Assess whether the EPA analysis appropriately incorporates popula-

tion differences, including diet, when extrapolating from the Taiwanese study population to the U.S. population.

• Evaluate whether the dose-response analysis conducted by EPA and any other available analyses of more recent data are adequate for estimating an effective dose for a 1% response (ED_{01}).

• Determine whether EPA's analysis appropriately considers and characterizes the available data on the mode of action of arsenic and the information on dose-response relationships and uncertainties when assessing the public-health impacts.

• Determine whether EPA's risk estimates at 3, 5, 10, and 20 µg/L are consistent with available scientific information, including information from new studies.

THE SUBCOMMITTEE'S APPROACH TO ITS CHARGE

The subcommittee considered several hundred new scientific articles on arsenic published since the 1999 NRC report. It also heard presentations from the EPA administrator; other EPA representatives; the EPA Science Advisory Board; other scientists with expertise in arsenic toxicity; federal, state, and local government agencies; trade organizations; public-interest groups; and concerned individuals.

The subcommittee evaluated the arsenic hazard assessment conducted by EPA for the pending arsenic standard published in the January 22, 2001, *Federal Register* and considered the comments made in the EPA Science Advisory Board's December 2000 report on the previously proposed rule. The subcommittee was not asked to assess U.S. population exposures. It addressed scientific issues concerning the hazards from consumption of drinking water contaminated with arsenic. It did not comment or make recommendations on risk management or policy decisions. By definition, determining an MCL requires policy considerations, including risk-management options and cost-benefit analyses, which are beyond the scope of the charge to this subcommittee.

It should also be noted that the NRC was charged with updating the 1999 report *Arsenic in Drinking Water*, not with reviewing its own report. Therefore, the subcommittee has taken that report as a starting point in its evaluation of more recent information.

THE SUBCOMMITTEE'S EVALUATION

Epidemiological (Human) Studies

The 1999 NRC report concluded that arsenic is associated with both cancer and noncancer effects. At that time, there was sufficient evidence to conclude that ingestion of arsenic in drinking water causes skin, bladder, and lung cancer. The internal cancers (bladder and lung) were considered to be the main cancers of concern, and there was sufficient evidence from large epidemiology studies in southwestern Taiwan of a dose-response relationship between those cancers and exposure to arsenic in drinking water.

Since the publication of the 1999 report, evidence has increased that chronic exposure to arsenic in drinking water might also be associated with an increased risk of high blood pressure and diabetes. Pending further research that characterizes the dose-response relationship for high blood pressure and diabetes, the magnitude of possible risk that exists at low levels is not quantifiable. Nevertheless, even small increases in relative risk for these conditions at low dose could be of considerable public-health importance. This potential impact should be qualitatively considered in the risk-assessment process. Some evidence also published since the 1999 NRC report shows an association between arsenic ingestion and potentially adverse reproductive outcomes and noncancer respiratory effects. However, those data require confirmation.

Four major epidemiological studies have been published since the 1999 NRC report in which the association between internal cancers and arsenic ingestion in drinking water has been investigated. The data from three of those studies (one in Chile, one in northeastern Taiwan, and one in southwestern Taiwan) confirm the association between internal cancers and arsenic exposure through drinking water. Another study (in Utah) did not demonstrate such an association.

The strengths of the recent studies from Chile and northeastern Taiwan include the evaluation of some potential confounding factors affecting the observed association between arsenic ingestion and cancer in newly diagnosed cases. Although the recent study in southwestern Taiwan is limited in its exposure assessment, it addresses the issue of lifestyle differences (e.g., diet, smoking) that might have influenced mortality rates in the area where arsenic is endemic. In that study, cancer rates in the area of southwestern Taiwan where arsenic is endemic were compared with cancer rates in counties neighboring the area (where the lifestyle is similar) and with rates for all of Taiwan. The arsenic-related risk estimates based on the two different comparison populations did not differ substantially, indicating that lifestyle differences between the region of southwestern Taiwan where arsenic is endemic and the rest of Taiwan do not substantially affect estimates of the risk of cancers from ingesting arsenic in drinking water.

The study in Utah was the first large-scale study to attempt to consider the association between internal cancers (bladder and lung) and arsenic exposure through drinking water in a U.S. population. However, the subcommittee concluded that the limitations of the Utah study currently preclude its use in a quantitative risk assessment. One limitation was the unconventional method used in that study to characterize exposure. Furthermore, in contrast to the southwestern Taiwan study where lifestyle differences do not appear to influence relative risk of cancer from arsenic in drinking water, the Utah study used a comparison group with differences in lifestyle characteristics from the study population. The study population was composed of individuals with religious prohibitions against smoking, and the unexposed comparison group was the overall population of Utah, where such religious prohibitions are not practiced by all residents.

The other recent studies of arsenic in humans, taken together with the many studies discussed in the 1999 NRC report, provide a sound and sufficient database showing an association between bladder and lung cancers and chronic arsenic exposure in drinking water, and they provide a basis for quantitative risk assessment. The subcommittee concludes that the early data from southwestern Taiwan remain appropriate for use in dose-response assessment of arsenic in drinking water. In addition, recent studies increase the weight of evidence for an association between internal cancers and arsenic exposure through drinking water. In particular, data from northern Chile on risk of lung cancer incidence are also appropriate for use in a quantitative risk assessment.

Metabolism and Mode-of-Action Studies

When evaluating the hazards from arsenic in drinking water, it is important to evaluate data on the fate of arsenic in the body (i.e., its metabolism) and how it causes its adverse effects (i.e., its mode of action). Arsenic is metabolized in the body by reduction and methylation reactions. The main product of those reactions, dimethylarsinic acid, is readily excreted from the body in the urine, but recent data indicate that reactive and toxic intermediate metabolites may be distributed to tissues and excreted in urine. The mechanisms responsible for the adverse effects associated with arsenic, including some types of cancer, cardiovascular disease, and diabetes, probably occur through multiple independent and interdependent mechanisms. The shape of the dose-response curve for one type of adverse effect might have little relevance to the shape for a different effect. Likewise, the shape of the dose-response curve for disruption of a specific biochemical pathway by arsenic is not necessarily relevant to the overall shape of the dose-response curve for a complex disease process, such as tumor development following chronic exposure.

Biostatistical approaches are required in a dose-response assessment to extrapolate from the lowest concentrations of arsenic at which increases in cancer are observed in a study population to lower concentrations to which the study population of interest is exposed. The mode of action by which a chemical causes cancer can sometimes determine how human or animal data should be extrapolated and used to evaluate allowable drinking-water contaminant concentrations. If an agent acts directly to cause DNA damage, it is standard practice for the estimated risk of cancer to be extrapolated in a *linear* fashion from the lowest measured exposure to zero (i.e., below the range of observations, risk is assumed to be directly proportional to the exposure.) If an agent acts indirectly, the possibility of *sublinear* extrapolation is considered (i.e., such extrapolation has sometimes been interpreted to indicate a "threshold" for effects.) In the absence of definitive mode-of-action data, EPA's general policy is to use a linear extrapolation from the observed data range for its carcinogenic risk assessments. After concluding that the mode-of-action data were inadequate to define the shape of the curve, EPA made a policy-based decision to use a default assumption of linearity.

Although a large amount of research is available on arsenic's mode of action, the exact nature of the carcinogenic action of arsenic is not yet clear. Therefore, the subcommittee concludes that the available mode-of-action data on arsenic do not provide a biological basis for using either a linear or nonlinear extrapolation. Furthermore, in laboratory studies, cellular effects of arsenic occur at concentrations below those found in the urine of people who had ingested drinking water with arsenic at concentrations as low as 10 µg/L. Therefore, even if the curve is sublinear at some point (e.g., if a threshold exists), the available data showing cellular effects at arsenic concentrations in the range of those measured in U.S. populations suggest that any hypothetical threshold would likely occur below concentrations that are relevant to U.S. populations.

Variability and Uncertainty in an Arsenic Risk Assessment

Variability (differences in outcomes due to factors contributing to risk) and uncertainty (resulting from lack of knowledge in the underlying science) should be considered in an arsenic risk assessment. Differences in the exposures of individuals and populations and differences in responses to a given exposure result in variability in a response. Often, that variability can be measured and quantified, but in many cases, assumptions must be made about many of the variables when information is lacking.

Sources of variability in an arsenic risk assessment include exposure differences in subpopulations (e.g., infants and children), and variability in arsenic metabolism. Individual exposures to arsenic can be affected by a number of factors, particularly the variability in the amount of arsenic in drinking water, water-ingestion rates, arsenic content in different foods, food-consumption rates, and other characteristics of the exposed population, such as sex, age, and body weight. EPA made assumptions with regard to intake of drinking water (including that for cooking) and arsenic through food to account for difference between southwestern Taiwan and the United States when estimating its risks. The basis for those assumptions, however, is not clear and adds to the uncertainty in the risk estimates.

It has been argued that poor nutrition might make the Taiwanese population more susceptible to the effects of arsenic than the U.S. population and that generalizing from the Taiwan population to other populations with different diets and, possibly, nutritional status is inappropriate. However, the subcommittee concludes that there is no evidence of nutritional factors that could account for the high rate of cancer seen in the arsenic-exposed Taiwanese population. Furthermore, similar increases in risk have been associated with chronic arsenic exposure in many other countries, including Chile and Argentina, where poor nutrition and low-protein diets are not issues. Therefore, the subcommittee concludes that the risk estimates based on the southwestern Taiwanese data are not substantially affected by differences in nutritional status or diet.

The subcommittee evaluated data to determine whether there is evidence that infants and children are more susceptible than adults to the effects of arsenic. There are no reliable data that indicate heightened susceptibility of children to arsenic. The subcommittee agrees that infants and children might be at greater risk for cancer and noncancer effects because of greater water consumption on a body-weight basis. However, cancer remains the health end point of concern, and the lifetime cancer risk estimates account for the greater childhood exposures by deriving risk estimates from epidemiology studies of cancer among populations exposed to arsenic since birth, as was the case for most of the populations in which the association between arsenic and cancer was studied.

Considerable variability in metabolism of arsenic in humans is reflected, in part, by differences in the pattern of excreted arsenic metabolites in the urine. Because arsenic metabolites differ in their toxicity, variation in the metabolism of arsenic is likely to be associated with variations in susceptibility to arsenic. Genetic factors, age, the dose of arsenic received, and simultaneous exposure to other compounds, such as micronutrients, appear to be important considerations in arsenic metabolism. The fact that the metabolism of arsenic varies markedly between individuals should be considered in an arsenic risk assessment; however, at the present time it is uncertain how to account for that variability in a quantitative dose-response analysis.

The method used to characterize arsenic dose in a study is a source of uncertainty in arsenic dose-response assessment. The measurement of dose (e.g., cumulative exposure, lifetime average exposure, or peak exposure) that is most closely correlated with cancer outcomes is not well established. If an incorrect measurement of dose is used, then the relationship between dose and effect might be obscured. The choice of the dose measurement affects the interpretation of an epidemiological study and the choice of the dose-response model.

Smoking is a well-recognized risk factor for lung and bladder cancer, the two internal cancers mostly strongly associated with arsenic ingestion. There are no data available to indicate that smoking is a significant confounder of the observed association between exposure to arsenic in drinking water and an increase in lung or bladder cancer. However, several of the epidemiological studies reviewed by the subcommittee suggest the possibility of an interaction between smoking and arsenic on the risk of lung cancer or bladder cancer, but this potential effect requires further confirmation and characterization. If an interaction between smoking and arsenic were to exist, then differences in smoking prevalence between populations might influence the impact of using relative risks from one population to derive risk estimates in another population. The direction of this impact could be in either direction, that is, it could theoretically either increase or decrease the risk estimates, depending on the relative smoking prevalences.

Quantitative Evaluation of Arsenic Toxicity

For the southwestern Taiwanese study, risks can be estimated either by comparing cancer mortality in the human study population exposed to arsenic with cancer mortality in the general Taiwanese or the regional population (i.e., a mostly unexposed external comparison group) or by making comparisons within the study group between high- and low-exposed individuals (i.e., internal comparison group). The approach of using an unexposed external comparison population is classically used in the analysis of data similar to those available from Taiwan and has the advantage of minimizing exposure misclassification (e.g., classifying low-exposed individuals in the study population as unexposed). A potential disadvantage of using an external comparison group is that the analysis can be biased if the study population differs from the comparison population in important ways. Because of concerns about differences between the unexposed external comparison population and the study population in southwestern Taiwan, EPA used an internal comparison population in its dose-response assessment. As discussed above, however, results of a recent study in southwestern Taiwan indicate that differences in lifestyle factors between the region of southwestern Taiwan where arsenic is endemic and the rest of Taiwan do not appear to affect the risk of cancer from arsenic in drinking water. Therefore, the subcommittee derived its estimates of cancer risk by comparing the arsenic-exposed southwestern Taiwanese population with an external population, and it recommends that approach for arsenic risk assessments.

The subcommittee estimated ED_{01} values (i.e., the exposure dose at which there is a 1% response in the study population) for various studies using several different types of statistical models. The estimated ED_{01} values from the Chilean study on lung cancer ranged from 5 to 27 µg/L, depending on the exposure data used. The ED_{01} values estimated for the southwestern Taiwanese study ranged from 33 to 94 µg/L for lung cancer, and from 102 to 443 µg/L for bladder cancer, depending on the choice of statistical model. The previous NRC Subcommittee on Arsenic in Drinking Water estimated ED_{01} values for male bladder cancer mortality of 404 to 450 µg/L, depending on the model used. Those values are approximately within the range of ED_{01} values estimated by this subcommittee. However, because the ED_{01} values reported by the previous and current NRC subcommittees were derived through different biostatistical approaches, they are not directly comparable. The ED_{01} values in the 1999 NRC report reflect a 1% increase relative to background cancer mortality in Taiwan, whereas the current subcommittee's approach reports ED_{01} values based on a 1% increase relative to the background cancer mortality in the United States. The differences between these two approaches are discussed in a later section.

The subcommittee investigated the extent of the variability among different types of statistical models using a model-weighting approach and also assessed the impact of differences in background incidence rates between different populations when using relative risks in a risk assessment. In addition, statistical analyses were conducted to investigate the sensitivity of the resulting risk estimates to differences in water intakes and measurement error.

Research Needs

More research is needed on the possible association between arsenic exposure and cancers other than skin, bladder, and lung, as well as noncancer effects, particularly impacts on the circulatory system (high blood pressure, heart disease, and stroke), diabetes, and reproductive outcomes. Future studies of the relationships between arsenic ingestion and both noncancer and cancer outcomes should be designed to have sufficient power to determine risks in potentially susceptible subpopulations, including children; they should consider factors (e.g., smoking, diet, genetics) that could influence susceptibility to arsenic; and they should collect detailed exposure information, all in an effort to reduce uncertainty in the risk assessment. In addition, more information is needed on the variability in metabolism of arsenic among individuals and the effect of that variability on an arsenic risk assessment. Laboratory and clinical research is also needed to define the mechanisms by which arsenic induces cancer to clarify the risks at lower doses.

OVERALL CONCLUSIONS

There is a sound database on the carcinogenic effects of arsenic in humans that is adequate for the purposes of a risk assessment. The subcommittee concludes that arsenic-induced internal (lung and bladder) cancers should continue to be the principal focus of arsenic risk assessment for regulatory decision making, as discussed and as recommended in the 1999 NRC report. The human data from southwestern Taiwan used by EPA in its risk assessment remain the most appropriate for determining quantitative lifetime cancer risk estimates. Human data from more recent studies cited in this report, especially those from Chile, provide additional support for the risk assessment. In view of new data from southwestern Taiwan, the subcommittee recommends using an external comparison population, rather than high- and low-exposure groups within the exposed population, when analyzing the earlier studies from southwestern Taiwan. The observed data should be analyzed, using a model that is biologically plausible and provides a reasonable statistical fit to the data. For the southwestern Taiwanese cancer data, this model is the additive Poisson model with a linear term used for dose. The available data on the mode of action of arsenic do not indicate what form of extrapolation (linear or nonlinear) should be used below arsenic concentrations at which cancers have been observed in human studies. As discussed previously, there are no experimental data to indicate the concentration at which any theoretical threshold might exist. Therefore, the curve should be extrapolated linearly from the ED_{01} to determine risk estimates for the potential concentrations of concern (3, 5, 10, and 20 µg/L). The choice for the shape of the dose-response curve below the ED_{01} is, in part, a policy decision. It should be noted, however, that the Taiwanese and other human studies include data on exposures at arsenic concentrations relatively close to some U.S. exposures. Consequently, the extrapolation is over only a relatively small range of arsenic concentrations. The uncertainty associated with the assumptions in the analyses was discussed earlier.

The subcommittee's estimates of theoretical lifetime excess risk of lung cancer and bladder cancer for U.S. populations at different concentrations of arsenic in drinking water are presented in Table ES-1. These are maximum-likelihood (central-point) risk estimates, not upper-bound (worst-case) estimates.

Because a relative risk approach using data from Taiwan and Chile was used to project risks in the U.S. population, differences in the background rate of the disease can have an important impact on the overall risk estimate. The background incidence of lung or bladder cancer in Taiwan is lower than that in the United States; therefore, the projected risk estimates for those cancers will also be lower in Taiwan than in the United States. The corresponding risks estimated using Taiwanese background cancer rates would be approximately 2-fold lower for female bladder cancer, 3-fold lower for male bladder cancer, 3-fold lower for female lung cancer, and 2-fold lower for male lung

TABLE S-1 Theoretical Maximum-Likelihood Estimates[a] of Excess Lifetime Risk (Incidence per 10,000 People) of Lung Cancer and Bladder Cancer for U.S. Populations Exposed at Various Concentrations of Arsenic in Drinking Water[b,c]

Arsenic Concentration (μg/L)	Bladder Cancer		Lung Cancer	
	Females	Males	Females	Males
3	4	7	5	4
5	6	11	9	7
10	12	23	18	14
20	24	45	36	27

[a] The maximum-likelihood estimate is the central point estimate from the distribution of risk calculated using a particular statistical model and data set (see note b).

[b] Estimates were calculated using data from individuals in the region of southwestern Taiwan where arsenic is endemic, data from an external comparison group from the overall southwestern Taiwan area, and U.S. age-adjusted cancer incidence data. The risks are estimated using what the subcommittee considered reasonable assumptions: a U.S. resident weighs 70 kg, compared with 50 kg for the typical Taiwanese, and the typical Taiwanese drinks just over 2 liters of water per day, compared with 1 liter per day in the United States; therefore, it assumes that the Taiwanese exposure per kilogram of body weight is approximately 3 times that of the United States. It is possible to get higher and lower estimates using other assumptions. Risk estimates are rounded to the nearest integer. All 95% confidence limits are less than ±12% of the maximum-likelihood estimate and are not presented. Those confidence limits reflect statistical variability in the population incidence estimates only, a narrow range that primarily reflects the relatively large sample size of the data modeled. As such, they are not indicative of the true uncertainty associated with the estimates.

[c] If Taiwanese baseline cancer rates are used instead of U.S. data to estimate the risk, the corresponding risk estimates (incidence per 10,000) for arsenic at concentrations of 3, 5, 10, and 20 μg/L of drinking water are as follows: female bladder cancer, 2, 4, 8, and 15; male bladder cancer, 2, 3, 7, and 13; female lung cancer, 2, 3, 6, and 12; and male lung cancer, 2, 3, 6, and 11.

cancer (see Table ES-1, footnote c). It should be noted that standard epidemiological practices support the use of the background incidence rate in the country of interest when comparing relative risks across different populations. However, the subcommittee members are divided in opinion on whether using the U.S. background cancer incidence rate was preferable to using the Taiwanese background rate; some members of the subcommittee felt strongly that using the U.S. background rate was the preferred approach, while others felt that there was not sufficient justification to select one background rate over the other.

At a concentration of arsenic in drinking water of 3 μg/L, the subcommittee's theoretical lifetime risk estimates for bladder and lung cancer combined are between approximately 4 and 10 per 10,000 when risks are estimated using the Taiwan or U.S. background rates of these cancers, respectively. As discussed in Chapter 5, the subcommittee's risk estimates for lung cancer, based on the southwestern Taiwanese data and new analyses, are consistent with published risk estimates based on other data sets (e.g., Chile) and on other published analyses of the southwestern Taiwanese data. The estimates from this subcommittee are also generally consistent with the bladder cancer risk estimates presented in the 1999 NRC report. Risk estimates for lung cancer were not presented in the 1999 report.

EPA did not publish the theoretical risk estimates on which it based its analyses; its analyses were adjusted for the occurrence of arsenic in U.S. drinking water; such an analysis of arsenic concentrations in U.S. drinking-water supplies is beyond the charge to this subcommittee. Therefore, the subcommittee has compared its risk estimates to estimates calculated from the published analyses on which EPA based its risk estimates; those estimates were not adjusted for water consumption or arsenic in food in the same manner by EPA, nor by this subcommittee. The adjustments used by EPA for food and water consumption would have the effect of decreasing the risk estimates.

Even without those adjustments, the risk estimates on which EPA based its analyses are lower than this subcommittee's estimates. Several factors contribute to that difference. The subcommittee used an external comparison population, rather than an internal comparison as was done in EPA's analyses. The subcommittee also used a different statistical method from that used for the estimates on which EPA based its estimates of lifetime excess cancer risks. Also, the subcommittee has presented estimates based on both U.S. and Taiwanese background incidence data; EPA's estimates took into account only Taiwanese background incidence data. In addition, the method that the subcommittee used to adjust for arsenic in food and its assumptions regarding water intake in the United States and Taiwanese populations were different from those used by EPA in its analyses. These factors are summarized in Table 6-2.

As discussed in Chapter 6, even at the highest risk estimates made by the subcommittee, the increases in cancer due to arsenic in drinking water would be difficult to detect statistically in the U.S. population. For example, a lifetime excess risk of bladder cancer incidence in males of 45 per 10,000 would represent only 13% of the total risk for male bladder cancer in the United States from all causes. Epidemiological detection of such a risk would require study of a large population of individuals who consumed drinking water containing arsenic at a concentration of 20 µg/L over an extended period of time. Detection would be further complicated by variability in the concentrations of arsenic in drinking water, the unknown distribution of other risk factors (including smoking), and the mobility of the U.S. population. Because background lung cancer mortality in the United States is almost 10-fold greater than bladder cancer mortality, it would be even more difficult to demonstrate an association of arsenic in drinking water with lung cancer risk. Therefore, although the subcommittee's risk estimates are of public-health concern, they are not high enough to be detected easily in U.S. populations by comparing geographical differences in the rates of specific cancers with geographical differences in the levels of arsenic in drinking water.

In accordance with its charge, the subcommittee has not conducted an exposure assessment, subsequent risk characterization, or risk assessment. The theoretical lifetime excess cancer risks estimated by the subcommittee and the uncertainties surrounding those estimates as presented in this report should be interpreted in a public-health context that uses an appropriate risk-management framework.

In summary, the subcommittee concludes that recent studies and analyses enhance the confidence in risk estimates that suggest chronic arsenic exposure is associated with an increased incidence of bladder and lung cancer at arsenic concentrations in drinking water that are below the current MCL of 50 µg/L. The results of this subcommittee's assessment are consistent with the results presented in the NRC's 1999 *Arsenic in Drinking Water* report and suggest that the risks for bladder and lung cancer incidence are greater than the risk estimates on which EPA based its January 2001 pending rule.

1

Introduction

The U.S. Environmental Protection Agency (EPA) is required under the Safe Drinking Water Act (SDWA) to establish the concentrations of contaminants permitted in public drinking-water supplies. There are two drinking-water contaminant concentrations that the statute requires EPA to set—the maximum contaminant level goal (MCLG) and the maximum contaminant level (MCL). The MCLG is a health goal to be based on the best available, peer-reviewed scientific data. It is to be set at a concentration at which no known or anticipated adverse health effects occur, allowing for adequate margins of safety. The MCLG is not a regulatory requirement and might not be attainable with current technology or analytical methods. The MCL, in contrast, is an enforceable standard. The MCL is required to be set as close to the MCLG as is technologically feasible, taking cost into consideration.

The EPA Office of Water is considering options for revising the current MCL of 50 µg/L for arsenic in drinking water. To incorporate the most recent scientific research into its decision, in April 2001, the Office of Water requested that the National Research Council (NRC) review the data published on the health effects of arsenic since the NRC's 1999 report, *Arsenic in Drinking Water*. The NRC was also asked to evaluate the arsenic risk assessment conducted by EPA for the proposed arsenic standard published in the January 22, 2001, *Federal Register* (EPA 2001a).

SUMMARY OF THE
NRC'S 1999 *ARSENIC IN DRINKING WATER* REPORT

In 1996, the NRC Subcommittee on Arsenic in Drinking Water was form-ed and charged with reviewing EPA's 1988 risk assessment for arsenic, re-viewing the scientific data relevant to an arsenic risk assessment, and identify-ing data gaps. The subcommittee did not provide a formal risk assessment or recommend an MCL. The subcommittee concluded, on the basis of epidemio-logical studies in Taiwan, Chile, and Argentina, that "ingestion of arsenic in drinking water poses a hazard of cancer of the lung and bladder, in addition to cancer of the skin." The subcommittee further concluded, "No human studies of sufficient statistical power or scope have examined whether con-sumption of arsenic in drinking water at the current MCL (approximately 0.001 mg/kg (milligram per kilogram) body weight per day) results in an increased incidence of cancer or noncancer effects." Therefore, the subcom-mittee used the available epidemiological data and information on mode of action, uncertainty, and human susceptibility to extrapolate and estimate cancer risks at various concentrations of arsenic in drinking water. The sub-committee concluded that data from the area of southwestern Taiwan where arsenic is endemic provided the best data for dose-response assessment. Although data suggest an indirect mode of action, the subcommittee con-cluded that the data are insufficient to determine a threshold below which no effects are seen and are insufficient to depart from a linear extrapolation to zero. The subcommittee presented the risks using numerous models. Using the Poisson regression model, the lifetime risk for male bladder cancer at the MCL (50 micrograms per liter (µg/L)) is 1 to 1.5 per 1,000, and the lifetime risk for all cancers combined could be on the order of 1 in 100.[1] On the basis of its assessment, the subcommittee concluded that "the current EPA MCL for arsenic in drinking water of 50 µg/L does not achieve EPA's goal for public health protection and therefore requires downward revision as promptly as possible."

---

[1]Two of the 16 members of the previous subcommittee did not agree with the 1 in 100 estimate pending further analysis of the risk of lung cancer, as done for bladder cancer in Chapter 10 of the 1999 report.

POLICY BACKGROUND

In 1976, under the SDWA, EPA proposed an interim MCL of 50 μg/L for arsenic in drinking water as part of the National Interim Primary Drinking Water Standards; that standard was originally set in 1942 and would apply until EPA adopted a new MCL. In 1988, EPA conducted a risk assessment for arsenic in drinking water and, in 1996, requested that the NRC independently review the scientific database on the health effects of arsenic and evaluate the scientific validity of that risk assessment. *Arsenic in Drinking Water* (NRC 1999) was published in response to that request.

Summary of EPA's Actions

Following the publication of that report, EPA published a proposed arsenic MCL of 5 μg/L (EPA 2000a) in the *Federal Register*. That standard was based on the information in *Arsenic in Drinking Water* (NRC 1999) and on epidemiological studies from southwestern Taiwan, Chile, and Argentina, some of which were reviewed in the NRC report. EPA also reviewed data from an epidemiological study in Utah (Lewis et al. 1999) that was released after the NRC (1999) report. EPA concluded that "exposure to inorganic arsenic induces cancer in humans" and proposed a regulation on the basis of that effect. It further concluded that "[i]n the absence of a known mode of action(s), EPA has no basis for determining the shape of a sublinear dose-response curve for inorganic arsenic" (EPA 2000a) and, therefore, proposed using a linear type of a dose-response curve and setting an MCLG of zero. On the basis of "national and international research, the bladder cancer risk analysis provided by the National Research Council (NRC) report issued by the National Academy of Sciences (NRC 1999), and the NRC's qualitative statements of overall risk of combined cancers," EPA evaluated the risk posed by arsenic in drinking water at 3 μg/L, 5 μg/L, 10 μg/L, and 20 μg/L. According to EPA policy, an MCL is to be set "as close as feasible to the MCLG, based on available technology and taking costs to large systems into account" (EPA 2000a). EPA determined that 3 μg/L is technologically feasible but that the costs associated with compliance for that concentration are not justified by the benefits. Therefore, EPA proposed an MCL of 5 μg/L but requested public comment on alternative MCLs of 3 μg/L, 10 μg/L, and 20 μg/L. The effective date for compliance for small community water systems would be 5 years after the final rule issuance and 3 years for all other community water systems.

After the publication of the proposed rule, Morales et al. (2000) published a study in which a risk assessment for mortality from several internal cancers was presented. That risk assessment was based on reanalyses of the data from southwestern Taiwan (Chen et al. 1985). Risk estimates were calculated for mortality from lung, bladder, and liver cancers, as well as combined cancer deaths, using 10 different statistical models. Those models included linear, logarithmic and square-root models, calculated with and without a Taiwanese comparison population. EPA was considering those analyses in the final rule-making and, therefore, published a Notice of Data Availability in the *Federal Register* (EPA 2000b) summarizing and further analyzing the information from Morales et al. (2000).

On December 12, 2000, the EPA Science Advisory Board (SAB) issued a report on the proposed drinking-water regulation (EPA 2000c). In its majority report, the SAB commented on the scientific basis of EPA's health risk assessment and on the economic and engineering aspects of the final rule. Exposure assessment for the U.S. population, costs, benefits, control technologies, and policy issues discussed by the SAB go beyond the charge to this subcommittee and will not be discussed here. The SAB also discussed the differential sensitivity of children to arsenic but could not reach consensus on whether children have an increased sensitivity to arsenic. A minority report was written on that issue. In general, the SAB agreed with the analyses conducted by the NRC Subcommittee on Arsenic in Drinking Water (NRC 1999), which formed part of the basis for EPA's proposed regulation. It did not agree, however, with EPA's interpretation of the NRC report, stating that EPA "may have taken the modeling activity in the NRC report as prescriptive." The SAB also stated that EPA did not adequately address the NRC subcommittee's cautions about such factors as nutrition and exposure measures that were to be considered when using the Taiwanese data for assessing risk in the U.S. population. The SAB also concluded "that the comparison populations [used in some analyses] were not appropriate control groups for the study area" in the Taiwanese study, and, therefore, dose-response models that did not use the comparison population are more appropriate. The SAB did not believe, however, that "resolution of all these factors can nor must be accomplished before EPA promulgates a final arsenic rule." The SAB agreed that the "available data do not yet meet EPA's new criteria for departing from linear extrapolation of cancer risk."

In response to public comment and the SAB report, on January 22, 2001, EPA issued a pending standard of 10 µg/L (EPA 2001a). To support this revised proposed standard, EPA reviewed the available data and quantitatively estimated risks using the epidemiological data from southwestern Taiwan (Chen et al. 1988, 1992; Wu et al. 1989) because, despite limitations, "the Taiwan epidemiological studies provide the basis for assessing potential risk from lower concentrations of inorganic arsenic in drinking water, without having to adjust for cross-species toxicity interpretation." In its ruling, EPA discussed the implications for extrapolating from the observed range of arsenic concentrations in the Taiwanese population to the risk in the U.S. population of (1) arsenic in food and via cooking, (2) uncertainty in exposure from the grouping of individuals with respect to exposure, (3) nutritional aspects of the Taiwanese population, and (4) a study in Utah, which did not find any excess bladder or lung cancer risk after exposure was taken into account. "[T]he best available science provides no alternative to use of a linear dose-response process for arsenic because a specific mode (or modes) of action has not been identified"; therefore, EPA used a Poisson model, with concentration entered as a linear term to estimate the dose that results in a 1% increased risk (i.e., 1% benchmark dose; BMD_{01}), and extrapolated from that point using a linear extrapolation. That analysis follows the EPA draft carcinogen risk assessment guidelines (EPA 1996). On the basis of that assumption and estimates of water intake, EPA calculated cancer risks at various MCL options. Those risks were calculated for bladder and lung cancer combined. As described in the proposed rule (EPA 2000a), when estimating a risk at a given MCL option, EPA assumed that water-treatment facilities will treat water to achieve a concentration that is 80% of the MCL and, therefore, calculates risks based on 80% of the MCL. EPA made that assumption because "water systems tend to treat below the MCL level in order to provide a margin of safety" (EPA 2000a). For example, when estimating population risks at an MCL of 10 µg/L, the actual average exposure after water treatment would be 8 µg/L. Therefore, the estimated risks presented by EPA at an MCL of 10 µg/L are the risks calculated for a drinking-water concentration of 8 µg/L. Such an adjustment to account for the difference between the MCL and the actual drinking-water concentrations was not done in the risk estimates of the previous NRC subcommittee. In addition, the NRC (1999) risk estimates are for male bladder cancer alone, not combined cancers as presented by EPA, and estimates calculated with and without a comparison population (i.e., background data) are presented in the NRC report (1999). Given those risks, what is known or not known about arsenic's mode-of-action, cost-benefit analyses, and policy considerations, EPA set an MCLG of 0 µg/L and an MCL of 10 µg/L. Further details on the models and parameters used by EPA to determine their risk estimates are presented in Chapter 5.

On March 23, 2001, EPA published a notice that delays the effective date of the arsenic rule pending further study (EPA 2001b). EPA's Office of Water requested that the NRC update its 1999 report as part of the further study.

CHARGE TO THE SUBCOMMITTEE

In response to EPA's request, the NRC assigned the project to the Committee on Toxicology (COT), which convened the Subcommittee to Update the 1999 *Arsenic in Drinking Water* Report. Members selected to serve on this subcommittee have expertise in toxicology, epidemiology, cellular and molecular toxicology, biostatistics and modeling, risk assessment, uncertainty analyses, medicine, and public health. Five members of this subcommittee also served on the earlier Subcommittee on Arsenic in Drinking Water. This subcommittee is charged with the task of preparing a report updating the scientific analyses, uncertainties, findings, and recommendations of the *Arsenic in Drinking Water* report (NRC 1999) on the basis of several new studies and analyses published since the 1999 report was released. Specifically, the charge to the subcommittee was to review relevant toxicological and health-effects studies published and relevant data developed since the 1999 NRC report, including the toxicological risk-related analyses performed by EPA in support of its regulatory decision-making for arsenic in drinking water. The subcommittee addressed only scientific topics relevant to toxicological risk and health effects of arsenic; it did not address questions of economics, cost-benefit assessment, control technology, or regulatory decision-making.

The subcommittee performed the following tasks in response to the charge:

- It determined whether data from the 1988, 1989, and 1992 Taiwanese studies remain the best data for dose-response assessment and risk estimation.
- It assessed whether the EPA analysis appropriately incorporates population differences, including diet, when extrapolating from the Taiwanese study population to the U.S. population.
- It evaluated whether the dose-response analysis conducted by EPA, as well as any available analyses of more recent data, is adequate for estimating an effective dose for a 1% response (ED_{01}).

• It determined whether EPA's analysis appropriately considers and characterizes the available data on the mode of action of arsenic and the information on dose-response and uncertainties, when assessing the public-health impacts.

• It determined whether EPA's risk estimates at 3, 5, 10, and 20 µg/L are consistent with available scientific information, including information from new studies.

This subcommittee was not asked to assess and did not assess U.S. population exposures. Therefore, it did not estimate the benefits or potential lives saved in the United States at each possible regulatory level. Consistent with the makeup of the subcommittee, it has addressed the scientific issues related to the hazards from arsenic in drinking water and has avoided commenting or making recommendations on risk-management or policy decisions to the extent possible. To that end, the subcommittee has not recommended an MCL, which, by definition, requires policy considerations that include evaluation of risk-management options and cost-benefit analyses that are beyond the scope of this subcommittee. If a particular analysis conducted by the subcommittee has policy or risk-management decisions inherent in it, the subcommittee presents the analysis as an example and indicates which parameter values are derived from policy decisions.

It should also be noted that the subcommittee is charged with updating *Arsenic in Drinking Water* (NRC 1999) and not with reviewing it. Therefore, the subcommittee has taken that report as a starting point and has evaluated the information published since the release of that report.

ORGANIZATION OF THIS REPORT

Subsequent to the release of *Arsenic in Drinking Water* (NRC 1999), numerous new studies and analyses have been published on the health effects of arsenic and related data, including epidemiological studies on cancer and noncancer effects, toxicokinetic and mode-of-action studies, and reports on arsenic risk assessments. The remainder of the report is organized into five chapters that review those areas. To provide context for the update, each chapter includes a brief summary of the information from the 1999 report. Chapter 2 discusses recent data on the health effects of arsenic in humans. Chapter 3 presents information published since the 1999 report on the toxicokinetics and possible mode of action of arsenic. Chapter 4 addresses variability and uncertainty, which are to be considered when assessing the hazards of arsenic, arising from interindividual variability, dietary and nutritional factors, exposure issues in epidemiology studies, and latency periods. Chapter 5 provides a discussion on the modeling approaches for quantitative evaluation of risks. It includes analyses conducted by EPA in its proposed regulations and the comments of the EPA Science Advisory Board (SAB) as well as analyses conducted by this subcommittee. Chapter 6 presents a summary of the subcommittee's findings, including lifetime cancer risk estimates associated with lifetime consumption of arsenic at different concentrations in drinking water.

REFERENCES

Chen, C.J., Y.C. Chuang, T.M. Lin, and H.Y. Wu. 1985. Malignant neoplasms among residents of a blackfoot disease-endemic area in Taiwan: High-arsenic artesian well water and cancers. Cancer Res. 45(11 Pt 2):5895-5899.

Chen, C.J., M. Wu, S.S. Lee, J.D. Wang, S.H. Cheng, and H.Y. Wu. 1988. Atherogenicity and carcinogenicity of high-arsenic artesian well water. Multiple risk factors and related malignant neoplasms of blackfoot disease. Arteriosclerosis 8(5):452-460.

Chen, C.J., C.W. Chen, M.M. Wu, and T.L. Kuo. 1992. Cancer potential in liver, lung, bladder and kidney due to ingested inorganic arsenic in drinking water. Br. J. Cancer 66(5):888-892.

EPA (U.S. Environmental Protection Agency). 1996. Carcinogenic Risk Assessment Guidelines. Notice. Fed. Regist. 61(79):17960-18011.

EPA (U.S. Environmental Protection Agency). 2000a. 40 CFR Parts 141 and 142. National Primary Drinking Water Regulations. Arsenic and Clarifications to Compliance and New Source Contaminants Monitoring. Notice of proposed rulemaking. Fed. Regist. 65(121):38887-38983.

EPA (U.S. Environmental Protection Agency). 2000b. 40 CFR Parts 141 and 142. National Primary Drinking Water Regulations. Arsenic and Clarifications to Compliance and New Source Contaminants Monitoring. Notice of data availability. Fed. Regist. 65(204):63027-63035.

EPA (U.S. Environmental Protection Agency). 2000c. Arsenic Proposed Drinking Water Regulation: A Science Advisory Board Review of Certain Elements of the Proposal, A Report by the EPA Science Advisory Board. EPA-SAB-DWC-01-001. Science Advisory Board, U.S. Environmental Protection Agency, Washington, DC. [Online]. Available: http://www.epa.gov/sab/fiscal01.htm .

EPA (U.S. Environmental Protection Agency). 2001a. 40 CFR Parts 9, 141 and 142. National Primary Drinking Water Regulations. Arsenic and Clarifications to

Compliance and New Source Contaminants Monitoring. Final Rule. Fed. Regist. 66(14): 6975-7066.

EPA (U.S. Environmental Protection Agency). 2001b. 40 CFR Parts 9, 141 and 142. National Primary Drinking Water Regulations. Arsenic and Clarifications to Compliance and New Source Contaminants Monitoring. Final rule. Delay of effective date. Fed. Regist. 66(57):16134-16135.

Lewis, D.R., J.W. Southwick, R. Ouellet-Hellstrom, J. Rench, and R.L. Calderon. 1999. Drinking water arsenic in Utah: A cohort mortality study. Environ. Health Perspect. 107(5):359-365.

Morales, K.H., L. Ryan, T.L. Kuo, M.M. Wu, and C.J. Chen. 2000. Risk of internal cancers from arsenic in drinking water. Environ. Health Perspect. 108(7):655-661.

NRC (National Research Council). 1999. Arsenic in Drinking Water. Washington, DC: National Academy Press.

Wu, M.M., T.L. Kuo, Y.H. Hwang, and C.J. Chen. 1989. Dose-response relation between arsenic concentration in well water and mortality from cancers and vascular diseases. Am. J. Epidemiol. 130(6):1123-1132.

2

Human Health Effects

This chapter discusses the health effects of arsenic observed in human studies. It begins with a summary of the 1999 NRC report *Arsenic in Drinking Water*. Following that, the noncancer and cancer studies published since the 1999 report are discussed.

SUMMARY OF HUMAN HEALTH EFFECTS DISCUSSED IN THE 1999 REPORT

The previous Subcommittee on Arsenic in Drinking Water reviewed the health effects seen in humans following exposure to inorganic arsenic in drinking water. The subcommittee concluded that the observed health effects were dependent on the dose and duration of exposure. Overt nonspecific gastrointestinal effects, such as diarrhea and cramping; hematological effects, including anemia and leukopenia; and peripheral neuropathy might occur after weeks or months of exposure to high doses of arsenic (0.04 mg/kg/day). These acute or subacute effects are typically reversible. Specific dermal effects are characteristic of chronic arsenic exposure. Diffuse or spotted hyperpigmentation has been seen after 6 months to 3 years by chronic ingestion of high doses of arsenic (0.04 mg/kg/day, or 40 µg/kg/day) or 5 to 15 years of ingestion of low doses (on the order of 0.01 mg/kg/day or higher). Palmer-plantar hyperkeratosis is usually evident within years of the initial appearance of arsenical hyperpigmentation. Perturbed porphyrin metabolism and irreversible noncirrhotic portal hypertension have been seen following chronic exposure to 0.01 to 0.02 mg/kg/day or higher. Chronic exposure to doses sufficient to cause cutaneous effects has been associated with peripheral vascular disease in studies in Taiwan, Chile, northern Mexico, Japan, and Germany. A risk of mortality from hypertension and cardiovascular disease has also been associated with chronic exposure to arsenic. An association has been reported between chronic ingestion of arsenic in drinking water and an increased risk of diabetes mellitus. Some evidence also suggested that the ingestion of arsenic can have effects on the immune and respiratory systems. Teratogenic effects were seen following parenteral arsenic exposure in a number of mammalian species, but little evidence suggests that those effects follow oral or inhalation exposure. There were inadequate data to draw conclusions on the effects of arsenic on fertility and pregnancy outcomes.

Cancer had been seen following exposure to inorganic arsenic in drinking water. Ingestion of inorganic arsenic was an established cause of skin cancer at the time of the previous report (NRC 1999). On the basis of data from several epidemiological studies, particularly those examining exposed populations in Taiwan, Argentina, and Chile, the Subcommittee on Arsenic in Drinking Water concluded that the "evidence is now sufficient to include bladder and lung cancer among the cancers that can be caused by ingestion of inorganic arsenic" (NRC 1999). The subcommittee further concluded that although some evidence indicated an increased risk of cancers other than skin, lung, and bladder, the database was not as strong, and confirmatory studies would be needed to establish arsenic as an underlying cause in other cancers.

RECENT STUDIES OF NONCANCER EFFECTS IN HUMANS

In this section, recent studies of the reproductive, neurological, cardiovascular, respiratory, hepatic, hematological, diabetic, and dermal effects of arsenic are presented. No relevant studies were identified for other noncancer end points.

Cardiovascular Effects

Earlier epidemiological studies have indicated that the cardiovascular system might be sensitive to chronic ingestion of arsenic. Effects seen following chronic exposure to arsenic in drinking water include hypertension and increased cardiovascular-disease mortality (NRC 1999).

Rahman et al. (1999) conducted a cross-sectional evaluation of blood pressure in 1,595 adults (above 30 years of age) who resided their entire lives in one of four villages in a rural area of Bangladesh. Many villages in Bangladesh have high arsenic exposures resulting from the use of groundwater for drinking water; wells were drilled because of microbial contamination of surface waters. Well-water arsenic concentrations were determined by reference to a database compiled from recent surveys, and most contained arsenic concentrations in excess of 0.05 mg/L. Examiners were not completely blinded to a subject's general exposure status but had no knowledge of precise levels of exposure. No subjects were taking antihypertensive medication, and the diet, lifestyle, and socioeconomic status of all subjects were similar. Prevalence ratios for hypertension were assessed after adjustment for age, sex, and body-mass index. The Mantel-Haenszel-adjusted prevalence ratios for hypertension increased with increasing arsenic in drinking water. The ratios were 1.2 (95% confidence interval (CI) = 0.6-2.3; 50 cases, 573 controls), 2.2 (95% CI = 1.1-4.3; 93 cases, 483 controls), and 2.5 (95% CI = 1.2-4.9; 55 cases, 227 controls) for exposure category I (arsenic at <0.5 mg/L), II (0.5-1.0 mg/L), and III (>1.0 mg/L), respectively. The chi-square test for trend was highly significant ($p \ll 0.001$). Cumulative arsenic consumption (arsenic concentration of the well water multiplied by the years of consumption) was also analyzed. The adjusted prevalence ratios for hypertension were 0.8 (95% CI = 0.3-1.7; 13 cases, 225 controls); 1.5 (95% CI = 0.7-2.9; 83 cases, 610 controls), 2.2 (95% CI = 1.1-4.4; 40 cases, 239 controls), and 3.0 (95% CI = 1.5-5.8; 62 cases, 209 controls) for cumulative exposures of less than 1.0 mg/L-years, 1.0-5.0 mg/L-years, greater than 5.0-10.0 mg/L-years, and greater than 10.0 mg/L-years, respectively. The chi-square test for trend was highly significant ($p \ll 0.001$). In a linear regression model that took into account age, sex, and body-mass index, mean blood pressure increased with both exposure measures. This study is consistent with early reports in Taiwan associating average and cumulative arsenic exposure in drinking water with a risk of hypertension.

Lewis et al. (1999a) conducted a retrospective cohort mortality study of residents of Millard County, Utah, an area where some drinking-water wells contained concentrations of arsenic up to several hundred micrograms per liter. This study is critiqued in detail in the Cancer Effects section of this chapter. Relative to statewide rates, the cohort had an increased risk of death from "hypertensive heart disease" in males (standardized mortality ratio (SMR) = 2.20, 95% CI = 1.36-3.36) and in females (SMR = 1.73, 95% CI =

1.11-2.58). Although statistical power was limited, there was no evidence of a positive dose-response relationship. It is uncertain whether this finding reflects a chronic hypertensive effect of arsenic, because other causes of mortality more commonly associated with hypertension, particularly ischemic heart disease and cerebrovascular disease, were significantly decreased in the cohort. "Hypertensive heart disease" is infrequently encountered as a coded cause of death, and because the increased SMRs were based on small numbers (21 deaths in males and 24 deaths in females), only a slight degree of misclassification bias might have influenced the results. Deaths in the broad category of "nephritis and nephrosis" were increased in males (SMR = 1.72, 95% CI = 1.13-2.50), and the possibility exists that the grouping might have subsumed some cases of nonspecific nephrosclerosis associated with hypertension.

Hertz-Picciotto et al. (2000) conducted a reanalysis of circulatory disease mortality among a cohort of smelter workers to assess whether the healthy worker survivor effect (HWSE) might have obscured a potential contributory role of cumulative airborne arsenic exposure. Although the route of exposure might have been predominantly by inhalation, the study addresses an impact on a nonpulmonary systemic end point that might also have relevance to arsenic ingestion. Using the least-exposed cohort members as internal controls, analytical approaches that applied a time-lagging method for exposure assessment and adjusted for employment status as a time-dependent variable in each year of follow-up revealed a stronger association between arsenic exposure and cardiovascular disease mortality. A G-null approach that used time-period-dependent exposure rates rather than cumulative exposure to control for the HWSE found no relationship between arsenic and cardiovascular disease mortality; however, because of data-set limitations, the power of that analysis was low. The authors concluded that the HWSE might have contributed to the apparent lack of arsenic-associated cardiovascular disease mortality in prior occupational cohort studies, in contrast to positive effects observed in several drinking-water studies.

Dermal Effects

Chronic arsenic exposure causes a characteristic pattern of noncancer dermal effects that begins with spotted hyperpigmentation and might later include palmar and plantar hyperkeratosis. Many studies about those skin lesions in humans have been published (NRC 1999). Several recent studies have investigated adverse health effects associated with ingestion of arsenic present in groundwater in the Gangetic plane of West Bengal, India, and neighboring Bangladesh, where over the past decade more than 30 million people might have been consuming water with arsenic concentrations in excess of 50 μg/L (Chowdbury et al. 2000).

Mazumder et al. (1998) conducted a cross-sectional survey of the prevalence of truncal hyperpigmentation and palmar-plantar keratosis in a region of West Bengal, India, with groundwater arsenic concentrations ranging from nondetectable to 3,400 μg/L. Medical examinations, current volume of water consumption, and well-water arsenic measurements were obtained on 7,683 children and adults (4,093 females and 3,590 males) recruited from rural villages in a region known to have high groundwater arsenic concentrations and in a referent population from a region thought to have lower exposures (total population at risk was 150,457). Heterogeneity of exposure existed within each region, and examiners were blinded to the exact exposures of the subjects. For greater than 80% of the participants, well-water arsenic concentration was less than 500 μg/L. Age-adjusted prevalence of hyperpigmentation was 0.3/100 in females in the lowest exposure level (<50 μg/L) and increased to 11.5/100 in the highest exposure level (≥800 μg/L). Corresponding prevalences for males were 0.4/100 and 22.7/100. For keratosis, the prevalence for females ranged from zero in the lowest exposure level (<50 μg/L) to 8.5/100 in the highest exposure level (≥800 μg/L), and for males, it ranged from 0.2/100 to 10.7/100. Of note, 29 subjects with hyperpigmentation and 12 subjects with keratosis were consuming water from domestic wells containing arsenic concentrations less than 100 μg/L. However, as noted by the investigators, because of the cross-sectional nature of the investigation, the possibility existed that these individuals might have consumed water containing higher concentrations of arsenic at their worksites or at past residences.

A separate analysis (Mazumder et al. 1998) of 4,443 subjects examined the prevalence of hyperpigmentation and keratosis according to tertiles of arsenic dose calculated on a body-weight basis. The prevalence (not age adjusted) of hyperpigmentation was 0.0/100 for females and 0.4/100 for males in the lowest tertile (0 to 3.2 μg/kg/day); 3.0/100 for females and 6.5/100 for males in the middle tertile (3.2 to 14.9 μg/kg/day); and 5.9/100 for females and 16.6/100 for males in the highest tertile (14.9 to 73.9 μg/kg/day). One-tailed chi-square tests of trend were significant ($p < 0.001$) for males and females. The corresponding prevalences for keratosis in females were 0.7/100, 2.3/100, and 3.5/100 (test for trend $p = 0.028$) and for keratosis in males were 0.8/100, 4.2/100, and 12.7/100 (test for trend $p < 0.001$).

Tondel et al. (1999) conducted a cross-sectional survey for characteristic arsenic-related skin lesions in 1,481 subjects (903 males, and 578 females) at least 30 years of age who resided in four rural villages of Bangladesh. The villages were selected because they exhibited a range of arsenic concentration in the drinking water (nondetectable to 2,040 µg/L). Although specific information was collected on the water source of each individual, the daily volume of water consumed was not ascertained, and examiners were apparently not blinded to subjects' exposure levels. The crude prevalence of any arsenic-associated skin lesion (pigmentary changes or keratosis) was 29/100. Data were analyzed in five exposure categories. The age-adjusted prevalence of any arsenic-related skin lesion increased in relation to the arsenic concentration of the drinking water. In males in the lowest exposure category (drinking-water arsenic, ≤ 150 µg/L), the age-adjusted prevalence of any arsenic-associated skin lesion was 18.6/100 and increased to 37.0/100 in the highest exposure category (>1,000 µg/L) (chi-square dose for trend, $p < 0.001$). The corresponding rates for females were 17.9/100 in the lowest exposure category and 24.9 in the highest exposure category (test for trend, $p < 0.02$). At similar levels of arsenic in drinking water, males had a higher prevalence of skin lesions than females. A dose-index, calculated as the arsenic concentration of a subject's drinking water divided by his or her body weight, was used to examine the trend in age-adjusted prevalence of skin lesions across three categories (≤ 5 µg/L/kg, >5 -10 µg/L/kg, and >10 µg/L/ kg). The corresponding age-adjusted prevalences for males were 19.6/100, 30.2/100, and 34.8/100 (test for trend, $p < 0.001$); for females, the respective rates were 19.7/100, 22.1/100, and 30.8/100 (test for trend, $p < 0.001$).

Ahsan et al. (2000) conducted a cross-sectional survey of "melanosis" (hyperpigmentation) and keratosis in three contiguous rural villages in Bangladesh within a region suspected of having elevated arsenic in groundwater. In the study, 87 males and 80 females were drawn from a total population of 300 residents; the investigators indicated that there might have been a greater number of affected individuals volunteering than unaffected individuals, resulting in an overestimate of actual prevalence rates. Exposure variables included the concentration of total arsenic in a spot urine sample, the arsenic concentration of drinking water contained in storage pitchers found in each subject's residence, and a cumulative exposure index calculated by multiplying the arsenic concentration in the pitcher water by the estimated amount of water consumed per year and by the estimated number of years that the current tube-well had been used for drinking. Nine of the 167 participants had melanosis, 2 had keratosis, and 25 had both melanosis and keratosis. A surprising finding in this study was the relatively high prevalence of skin lesions in subjects whose current drinking-water samples contained very low concentrations of arsenic; 13.9% of all subjects with skin lesions were currently drinking water with an arsenic concentration of <10 µg/L. In the analyses that evaluated skin lesions as a function of cumulative arsenic index in milligrams, 7 of 38 subjects (18.4%) in the lowest quartile (<116.4 mg) had skin lesions. In the highest quartile (1,279.9-22,147.1 mg), 14 of 39 subjects (35.9%) had skin lesions. Compared with subjects in the lowest quartile of total urinary arsenic excretion, subjects in the highest quartile had an odds ratio of 3.6 for the presence of any skin lesion (95% CI = 1.2, 12.1; urinary arsenic at 471-1,840 µg/L). However, the correlation between creatinine-adjusted total urinary arsenic and arsenic concentration in drinking water was only 0.5.

Tucker et al. (2001) examined the relationship between skin lesions and arsenic ingestion in a cross-sectional study of 3,228 children and adults exposed to arsenic in drinking water in Inner Mongolia. Design aspects of this study are critiqued in the Cancer Effects section (see below). In an analytical approach that assessed exposure by using the "peak arsenic concentration" of the well water that was used during the subject's lifetime, there was a dose-dependent increase in the age-adjusted prevalence of both keratoses and "dyspigmentation" (truncal hyperpigmentation). Thirty-seven of the 172 subjects with keratoses and 25 of the 121 subjects with dyspigmentation were assigned peak arsenic concentration exposures of less than 100 µg/L.

Reproductive and Developmental Effects

Arsenic exposures have been associated with a number of adverse health outcomes. Relatively little attention, however, has been directed toward assessing the potential impact of arsenic on human reproductive health effects, despite studies in both humans and experimental animals demonstrating that arsenic and its methylated metabolites easily pass the placenta (Concha et al. 1998). Evidence from human studies suggests the potential for adverse effects on several reproductive end points. A small number of epidemiological studies investigating the relationship between arsenic exposure in humans and adverse reproductive effects have been published since the 1999 report, but they suffer from some limitations and are not necessarily applicable in populations exposed to arsenic in drinking water.

A hospital case-control study found an increase of stillbirths in relation to the proximity of maternal residence to an arsenical pesticide production plant in Texas (Ihrig et al. 1998). The study was small, however, and the findings seem to be restricted to specific subgroups. In addition, exposures to other agents from the chemical plants were not assessed.

An increase in infant mortality (divided into three subcategories: still-births, neonatal, and postneonatal) was observed in a county in northern Chile during a time when there was a substantial increase in the arsenic concentration of the public drinking-water supply (from 90 to 800 µg/L) (Hopenhayn-Rich et al. 2000). A subsequent decrease in arsenic was accompanied by a return to the normal Chilean trend in infant mortality over time.

A retrospective survey in Bangladesh (Ahmad et al. 2001) compared several outcomes in women exposed to high (mean 240 µg/L) and low (below 20 µg/L) arsenic concentrations in drinking water and found increases in spontaneous abortions, stillbirths, and preterm births. This study was based on recall of previous pregnancies, however, and ascertainment of the outcomes was not clearly defined.

In summary, although several studies have addressed the potential reproductive effects of arsenic exposures in humans, the evidence is not conclusive. Many of the studies lacked information on lifestyle or personal factors that affect birth weight, congenital malformation, and other outcomes and information on other potential exposures. Several studies are ecological in nature and therefore subject to additional potential biases.

Neurological Effects

Previous studies of arsenic's neurological effects have generally focused on central-nervous-system effects seen following acute, high-dose intoxication and on the peripheral neuropathy that occurs following subacute or chronic exposure (NRC 1999). Two recent studies discussed below have focused on subtle cognitive effects in children following chronic exposure to arsenic.

Siripitayakunkit et al. (1999) investigated the association between environmental arsenic exposure and the intelligence of children in the Ronpiboon district of Thailand, where shallow artesian wells are contaminated with arsenic at concentrations as high as several milligrams per liter. Head-hair arsenic concentration and performance on the Weschler Intelligence Scale Test for Children (WISTC) in 529 schoolchildren (6 to 9 years of age) were analyzed in a cross-sectional evaluation. Median head-hair arsenic concentration was 2.42 µg/g (from 0.48 to 26.94 µg/g). Only 8.3% of the subjects had head-hair arsenic concentrations of 1 µg/g or less. Head-hair arsenic concentration was inversely associated with full-scale intelligence quotient (IQ) in an analysis adjusted for age, father's occupation, maternal intelligence, and family income. There were no data on exposure to lead, a major potential confounder, or on nutritional factors, such as iron. Although subjects lived in areas that apparently had different environmental arsenic concentrations, it was not stated whether the examiners were blinded to the location of residence. In addition, no data were presented regarding the concentration of arsenic in the children's drinking water, the concentration in the subjects' urine, or the presence of arsenic-related skin lesions in any subject. These factors limit the interpretation of the association between IQ and head-hair arsenic concentration.

A recent cross-sectional study in San Luis Potosi, Mexico (Calderon et al. 2001) examined the impact of arsenic and lead on the neuropsychological performance of schoolchildren aged 6 to 9 years. Subjects included 41 children living within 1.5 kilometers (km) of a smelter complex (Morales zone) with increased arsenic and lead concentrations and 39 children living 7 km upwind from the smelter (Martinez zone). The geometric mean total arsenic concentration in urine was higher in the Morales children (62.91 μg/g of creatinine; from 27.54 to 186.21) than in the Martinez children (40.28 μg/g of creatinine; from 18.20 to 70.79) ($p < 0.05$). Blood lead concentrations were similar in the two groups. Maternal and paternal educational attainment, socioeconomic status, and iron status were lower in the Martinez group. Neuropsychological performance was assessed using the Weschler Intelligence Scale for Children, Revised Version, for Mexico (WISC-RM). In a comparison unadjusted for indices of metal exposure or other covariates, the Morales children scored significantly higher than the Martinez children on the full-scale IQ test and other neuropsychological subscores. However, investigators reported an inverse correlation between log- transformed total urinary arsenic concentration (micrograms per gram of creatinine) and verbal IQ after adjustment by age, sex, socioeconomic status, maternal and paternal education, nutritional factors (transferrin saturation and height by age index), and blood lead. The relationship between log-transformed total urinary arsenic concentration and performance IQ was not significant. Speciation of arsenic in the urine (inorganic arsenic, monomethylarsonic acid (MMA), and dimethylarsinic acid (DMA)) was not performed, and the impact of past exposure to high levels of arsenic and lead emissions near the smelter was not assessed.

Respiratory Effects

Noncancer respiratory effects have been reported in populations exposed to arsenic in drinking water (NRC 1999), but the database on this topic has been sparse. Mazumder et al. (2000) recently reported an association between arsenic ingestion in drinking water and the prevalence of respiratory effects in a large cross-sectional survey of subjects residing in one of the arsenic-affected districts of West Bengal, India. Dermal effects have also been studied in this population (Mazumder et al. 1998). The analysis of the respiratory effects excluded 819 of the 7,683 individuals because of current or past history of smoking. Prevalence odds ratio (POR) estimates for abnormal chest sounds on physical examination were increased for subjects who had arsenic skin lesions and who consumed water with arsenic at greater than 500 μg/L, compared with subjects who had no skin lesions and who consumed water with arsenic at less than 50 μg/L. For females, the age-adjusted POR for cough was 7.8 (95% CI = 3.1-19.5) and for chest sounds, 9.6 (95% CI = 4.0-22.9). For males, the age-adjusted POR for cough was 5.0 (95% CI = 2.6-9.9) and for chest sounds, 6.9 (95% CI = 3.1-15.0). Examiners were not blinded to the presence of arsenic-related skin lesions. However, the large size of the study, the high odds ratios observed, and the positive trend with current arsenic exposure were such that the study provides support for reports of arsenic-associated pulmonary effects previously noted in Chile in the 1960s and 1970s (see NRC 1999).

Milton et al. (2001) reported an association between chronic arsenic ingestion and "chronic bronchitis" in a small cross-sectional study in Bangladesh of 94 individuals with arsenic-associated skin lesions. These individuals were attending "health awareness" meetings in three villages. The mean concentration of arsenic in drinking water was 614 μg/L, and the range was 136 to 1,000 μg/L. The study included 124 nonexposed individuals recruited from three villages "known to be not contaminated with arsenic." All participants never smoked, and all denied a history of asthma or tuberculosis. Chronic bronchitis was defined as a history of cough productive of sputum on most days for at least 3 consecutive months for more than 2 successive years combined with the presence of chest rhonchi and/or crepitations on physical examination. Chronic bronchitis was present in 14 of 40 exposed males, 11 of 50 nonexposed males, 15 of 54 exposed females, and 2 of 74 nonexposed females. The crude prevalence ratios for chronic bronchitis were 1.6 (95% CI = 0.8-3.1) and 10.3 (95% CI = 2.4-43.1) for males and females, respectively. After Mantel-Haenszel adjustment for sex, the prevalence ratio was 3.0 (95% CI = 1.6-5.3). Although this study could have considerable recruitment bias and observer bias, it contributes limited evidence to the studies that suggest an adverse respiratory effect of chronic high-dose arsenic ingestion.

In an ecological study of mortality in an area of southwestern Taiwan where blackfoot disease is endemic (Tsai et al. 1999; see Diabetes section for description), the SMR for "bronchitis" increased significantly relative to a nearby reference population (SMR = 1.53; 95% CI = 1.30-1.80), and to all of Taiwan (SMR = 1.95; 95% CI = 1.65-2.29). The SMR for emphysema was not significantly different from either reference population. The authors noted that it is unlikely that differences in the rate of smoking account for the increased bronchitis mortality.

Hepatotoxic Effects

Hernandez-Zavala et al. (1998) studied liver function in individuals from three towns in the Region Lagunera of Mexico. The mean arsenic concentration in drinking water in each village was 14.0 ± 3.1 µg/L, 116.0 ± 37 µg/L, and 239.0 ± 88 µg/L, and the corresponding mean total arsenic concentration in urine for each group (n = 17 per village) was 88.0 ± 27, 398.0 ± 258, and $2,058.0 \pm 833$ µg/L. The duration of exposure was not stated, but it was known that the middle-exposure town reduced the arsenic concentration (from 400 µg/L) in its water supply 3 years prior to the study. No subjects had recently consumed alcohol or had a history of chronic alcoholism. The mean concentrations of serum alkaline phosphatase and total bilirubin were significantly increased in individuals from the highest exposure town compared with the lowest- exposure town. Total urinary arsenic concentration was also correlated with those end points. Although data were not shown, the authors stated that those correlations were not significantly changed by adjusting for age, pesticide exposure, or history of alcohol or tobacco use or by examining separate correlations with urinary inorganic arsenic, MMA, or DMA. Serum transaminases (ALT, AST, and GGT) and albumin were normal and did not differ significantly between the groups. Twenty-six percent of the subjects from the middle- and high-exposure towns were reported to have at least one skin lesion associated with arsenic exposure. The potential impact of skin lesions as a covariate in the analysis was not reported. The impact of potential outliers on the results was not discussed. Although the correlations between urinary arsenic and both alkaline phosphatase and bilirubin concentrations observed in this study are interesting and might suggest an element of cholestasis, the biochemical findings noted are not specific for this hepatic end point. In the recent study by Santra et al. (1999), bilirubin or alkaline phosphatase increases were not a characteristic finding in 93 patients with firm hepatomegaly attributed to chronic arsenicosis in West Bengal, India. In that study, liver biopsy results from 69 patients with a clinical diagnosis of chronic arsenic poisoning revealed portal fibrosis in 63 (91.3%) cases, cirrhosis in 2 cases (2.9%), and normal histology in 4 (5.8%) cases. The degree of fibrosis was considered mild (grade 1) in 34 (53.9%) of the cases. Clinical evidence of portal hypertension (e.g., esophageal varices) was uncommon.

Hematological Effects

Hernandez-Zavala et al. (1999) also examined urinary porphyrin excretion and erythrocyte heme synthesis pathway enzymes in the same population described previously in the section Hepatotoxic Effects (Hernandez-Zavala et al. 1998). A dose- dependent increase in the ratio of urinary coproporphyrin III to coproporphyrn I and in the ratio of total coproporphyrin to total uroporphyrin was observed among the groups (17 individuals per group; mean urinary arsenic excretion, 88.0 ± 27, 398 ± 258, and $2,058 \pm 833$ µg/L). Those ratios are in contrast to earlier findings by those authors in a population with exposures that were somewhat higher and of longer duration (Garcia Vargas et al. 1994). In that study, chronic high-dose arsenic ingestion was associated with inversions of the ratio of coproporphyrin to uroporphyrin (i.e., ratio <1) and of coproporphyrin III to coproporphyrin I. It is not clear whether differences in the intensity and temporal pattern of arsenic dose contributed to the disparate findings in those studies.

Diabetes

The main focus of research looking at the effects of arsenic on the endocrine system is the association between arsenic exposure and diabetes mellitus. The NRC (1999) report reviewed studies in Taiwan and Bangladesh that associated chronic ingestion of arsenic in drinking water with an increased risk of diabetes mellitus.

More recently, as part of an ecological study examining multiple causes of mortality in a section of the area of southwestern Taiwan where blackfoot-disease is endemic, Tsai et al. (1999) examined mortality from diabetes mellitus in four townships where artesian well water containing arsenic (median concentration of 0.78 mg/L (780 µg/L); from 0.25 to 1.14 mg/L) had been consumed from the early 1900s until the mid- to late-1970s. Observed mortality between 1971 and 1994 was compared with age and sex-specific expected mortality based on data from (1) a local reference group derived from two nearby counties, and (2) all of Taiwan. The local reference group was considered to be similar to the study group with respect to lifestyle factors; however, the drinking-water arsenic concentration of the local reference area was not stated. The extended time of follow-up resulted in a relatively large number of deaths being available for analysis. Within the high arsenic area, there were 11,193 recorded deaths during 1,508,623 person-years of observation for males and 8,874 recorded deaths during 1,404,759 person-years of observation for females. Diabetes mellitus (World Health Organization's International Classification of Disease (ICD)(CDC 2001) 8 and 9, code 250) was listed as the underlying cause of death for 188 males and 343 females. For males, the SMRs were 1.35 (95% CI = 1.16-1.55) and 1.14 (95% CI = 0.98-1.31), using the local and the national reference group, respectively. For females, the corresponding SMRs were 1.55 (95% CI = 1.39-1.72) and 1.23 (95% CI = 1.11-1.37). Because only mortality attributed primarily to diabetes mellitus was examined, studies such as this one might underestimate an association between arsenic exposure and the population risk of the disease.

Tseng et al. (2000) recently reported the results of a prospective cohort study examining the incidence of diabetes mellitus in three villiages from the arsenic endemic area of southwestern Taiwan. The study population consisted of three villages where artesian well water (median arsenic concentration from 0.70 to 0.93 mg/L) was used for drinking and cooking until the mid-to-late 1970s. The study enrolled 632 subjects over 30 years of age who were determined not to have diabetes mellitus at the time the cohort was assembled (January and February, 1989). Details of the cohort are described by Lai et al. (1994). In 1991 and 1993, 446 of the cohort agreed to participate in a follow-up examination that included a fasting blood glucose and an oral glucose tolerance test. Diabetes mellitus was defined in accordance with the World Health Organization criteria. The incidence of diabetes mellitus in the study population was calculated as the total number of incident cases divided by the sum of follow-up person-time in all subjects. Data on each subject included age, sex, body-mass index, and an index of lifetime cumulative arsenic exposure (CAE). CAE (in units of milligrams per liter-years) was calculated as the product of the median arsenic concentration of the well water in every village that a subject inhabited at some point in his or her life multiplied by the length of time they consumed well water in that village. Individuals were excluded if complete arsenic exposure data were not available. Incidences for diabetes mellitus in the study population were compared with those reported for a demographically similar control population that was studied contemporaneously (Wang et al. 1997). Tseng et al. (2000) reported that during a follow-up period that included 1,499.5 person-years, 41 of 446 subjects developed diabetes mellitus (all noninsulin-dependent diabetes mellitus or type II diabetes mellitus). The incidence for new cases was particularly increased in subjects 55 years of age or older (50.8 per 1,000 person-years). The age-specific incidence density ratios were 3.6 (95% CI = 3.5-3.6), 2.3 (95% CI = 1.1-4.9), 4.3 (95% CI = 2.4-7.7), and 5.5 (95% CI = 2.2-13.5) for individuals 35-44, 45-54, 55-64, and 65-74 years of age, respectively. The relative risk for developing diabetes mellitus among those with more than 17 mg/L-years CAE compared with those with less than 17 mg/L-years CAE was 2.1 (95% CI = 1.1-4.2), adjusted for age, sex, and body-mass index in a multivariate Cox proportional hazards model. When considered as a continuous variable, the CAE was associated with an adjusted relative risk of developing diabetes mellitus of 1.03 for every 1 mg/L-years of exposure ($p < 0.05$).

RECENT STUDIES OF CANCER EFFECTS IN HUMANS

Since the previous evaluation of arsenic by NRC (1999), several studies have been completed that contribute to our understanding of dose-response relationships for arsenic in drinking water and cancer risk. In this section, detailed summaries and evaluations are presented of two of the recently completed studies that have adequate data to contribute to quantitative assessment of risk—one for urinary-tract cancers in Taiwan (Chiou et al. 2001) and the other for lung cancer in Chile (Ferreccio et al. 2000).

Other studies from Taiwan, Finland, and the United States with less information regarding risk of cancer of internal organs are also summarized (Tsai et al. 1998; Kurttio et al. 1999; Lewis et al. 1999a). For the most part, the limitations of those studies preclude direct application of their data to a quantitative risk assessment of arsenic in drinking water. However, one of the studies provides insight into issues surrounding the EPA risk assessment (Tsai et al. 1998). Two studies of skin cancer are also discussed (Karagas et al. 2001; Tucker et al. 2001). Skin cancers, however, are not as great a concern as internal cancers, such as bladder and lung cancer, because internal cancers are life-threatening, whereas most skin cancers are not. Skin cancer studies by Ma et al. (1999), Buchet and Lison (1998), Hinwood et al. (1999), Tsai et al. (1998), Ahmad et al. (1999), and Kurokawa et al. (2001) were noted by the subcommittee but are not discussed. Table 2-1 summarizes the major human studies in which cancer end points have been investigated. The table includes studies that were described in the 1999 NRC report as well as studies from the current evaluation.

Several qualitative criteria were used in evaluating the available epidemiological studies of cancer. It is important that studies adhere to basic epidemiological principles designed to avoid major sources of bias. In the case of ecological and cohort studies, these principles include accuracy of diagnoses (or cause of death), selection of an appropriate comparison population, and a clear definition of exposed and unexposed populations. In cohort studies, a high rate of successful follow-up is desirable. For case-control studies, evaluation criteria include careful attention to accurate diagnoses of cases, an adequate response rate among both cases and controls, and appropriate selection of the control group. In all studies, statistical power is a key consideration. Findings from small studies, even those with excellent methodology, are of limited utility.

The approach to exposure assessment and the use of estimated exposures in data analysis are key elements in epidemiological studies of all environmental exposures and were a central focus of the subcommittee's review of the literature of arsenic in drinking water and risk of cancer. Therefore, this section begins with a discussion of general issues in exposure assessment in epidemiological studies, followed by a discussion of specific studies.

General Issues with Exposure Measurement

In reviewing relevant data developed on measuring arsenic exposure since the 1999 NRC report *Arsenic and Drinking Water*, it is worthwhile to begin by considering concepts of exposure and dose, because numerous definitions and methods of estimating exposure have been developed. Exposure and dose are related but separate concepts. Exposure to chemicals occurs when there is a chemical source, transportation of the chemical from the source to the

TABLE 2-1 Bladder, Kidney, and Lung Cancer Mortality or Incidence in Epidemiological Studies on Arsenic

Study	Location	Exposure[a]	End Point	Cases, No.	Study Outcome	Comment
Ecological Studies Published Since *Arsenic in Drinking Water* (NRC 1999)						
Tsai et al. 1999	Blackfoot endemic area of SW Taiwan	Arsenic endemic area	Lung, liver, bladder, kidney and skin cancers (among many others)	Female Lung, 471	SMR: Female 4.13 (3.77-4.52) Regional 3.50 (3.19-3.84) National	SMRs calculated compared with mortality experience of the whole of Taiwan and compared with mortality in the two counties in SW Taiwan where the blackfoot endemic area is located
				Bladder, 295	14.07 (12.51-15.78) Regional 17.65 (5.70-19.79) National	
				Kidney, 128	8.89 (7.42-10.57) Regional 10.49 (8.75-12.47) National	
				Male Lung, 699	SMR: Male 3.10 (2.88-3.34) Regional 2.64 (2.45-2.84) National	
				Bladder, 312	8.92 (7.96-9.96) Regional 10.50 (9.37-11.73) National	
				Kidney, 94	6.76 (5.46-8.27) Regional 6.80 (5.49-8.32) National	
Ecological Studies Reviewed in *Arsenic in Drinking Water* (NRC 1999)						
Smith et al. 1998	Region II, Northern Chile	5-yr average, 420 µg/L average	Bladder cancer mortality	Male 93 Female 64	SMR: Male 6.0 (4.8-7.4) Female 8.2 (6.3-10.5)	Mortality, 1989-1993; national rates for 1991 used as the standard for the SMR; arsenic concentration is population-weighted average for major cities or towns in Region II, 1950-1974; information inadequate to calculate increase in risk per unit exposure
			Lung cancer mortality	Male 544 Female 154	SMR: Male 3.8 (3.5-4.1) Female 3.1 (2.7-3.7)	

Continued

TABLE 2-1 Continued

Study	Location	Exposure[a]	End Point	Cases, No.		Study Outcome	Comments
				Male	Female		
Hopenhayn-Rich et al. 1996, 1998	Cordoba Province, Argentina	County group:	Bladder cancer mortality			SMR:	Mortality, 1986-1991; national rates for 1989 used as the standard for the SMR; SMR for COPD below the expected level, indicating low smoking rates; also no trend with stomach cancer SMR
						Male Female	
		Low		113	39	0.80 (0.7-1.0) 1.21 (0.9-1.6)	
		Medium		93	24	1.42 (1.1-1.7) 1.58 (1.0-2.4)	
		High		131	27	2.14 (1.8-2.5) 1.82 (1.2-2.6)	
		County group:	Kidney cancer mortality			SMR:	
						Male Female	
		Low		66	38	0.87 1.00	
		Medium		66	34	1.33 1.36	
		High		53	27	1.57 1.81	
		County group:	Lung cancer mortality			SMR:	
						Male Female	
		Low		826	194	0.92 (0.85-0.98) 1.24 (1.06-1.42)	
		Medium		914	138	1.54 (1.44-1.64) 1.34 (1.12-1.58)	
		High		708	156	1.77 (1.63-1.90) 2.16 (1.83-2.52)	
Guo et al. 1997[b]	Taiwan	Data from 83,656 wells, national survey, towns grouped as follows:	Bladder cancer incidence (results shown are for transitional-cell carcinoma, the most common form of bladder cancer)	National rates		β(SE) from regression: Mixed results for exposure levels; at >0.64 ppm, the β(SE) was: Male Female 0.57 (0.07) 0.33 (0.04)	Incidence, 1980-1987
		<0.05 ppm 0.05-0.08 ppm 0.09-0.16 ppm 0.17-0.32 ppm 0.33-0.64 ppm >0.64 ppm	Kidney cancer incidence (results shown are for renal-cell carcinoma only)			β(SE) from regression: Mixed results for exposure levels; at >0.64 ppm, the β(SE) was: Male Female 0.03 (0.02) 0.142 (0.013)	

Reference	Location	Exposure / data	Endpoint	National data	Result	Comments
Chen and Wang 1990	Taiwan	1974-1976 data from 83,656 wells, national survey: average for each of 314 precincts or townships	Bladder cancer mortality	Male 23 Female 30	β(SE) from regression: Male 3.9 (0.5) Female 4.2 (0.5)	Mortality, 1972-1983 in 314 precincts and townships; regression-coefficient (β) estimates increase in age-adjusted mortality per 100,000 per 100 µg/L arsenic increase of water
			Kidney cancer mortality	Male 36 Female 36	β(SE) from regression: Male 1.1 (0.2) Female 1.7 (0.2)	
			Lung cancer mortality	Male 26 Female 30	β(SE) from regression: Male 5.3(0.9) Female 5.3(0.7)	
Wu et al. 1989	SW Taiwan	Average arsenic: <0.30 ppm 0.30-0.59 ppm ≥0.60 ppm	Bladder cancer mortality	Male 23 36 26 Female 30 36 30	Rate: Male 22.6 61.0 92.7 Female 25.6 57.0 111.3	Mortality, 1973-1986 in 42 villages in Taiwan
		Average arsenic: <0.30 ppm 0.30-0.59 ppm ≥0.60 ppm	Kidney cancer mortality	Male 9 11 6 Female 4 13 16	Rate: Male 8.42 18.90 25.26 Female 3.42 19.42 57.98	
		Average arsenic: <0.30 ppm 0.30-0.59 ppm ≥0.60 ppm	Lung cancer mortality	Male 53 62 32 Female 43 40 38	Rate: Male 49.16 100.67 104.08 Female 36.71 60.82 122.16	
Chen et al. 1985	SW Taiwan	Blackfoot-disease endemic area	Kidney cancer mortality	Male 42 Female 62	SMR: Male 7.72 Female 11.19	SMRs; age-specific Taiwan rates as standard
			Bladder cancer mortality	Male 167 Female 165	SMR: Male 11.00 (9.3-12.7) Female 20.09 (17.0-23.2)	
			Lung cancer morality	Male 332 Female 233	SMR: Male 3.20 Female 4.13	

Continued

TABLE 2-1 Continued

Cohort Studies Published Since *Arsenic in Drinking Water* (NRC 1999)

Study	Location	Exposure[a]	End Point	Cases, No.		Study Outcome		Comments
Chiou et al. 2001	NE Taiwan	Concentration of arsenic in household well water:	Urinary tract cancer incidence	Incidence rate (number of cases)		Relative risk:		Prospective cohort study of 8,102 persons; incidences calculated per 100,000; relative risks calculated using subjects with exposures ≤10 µg/L as a referent population
		≤10 µg/L		37.6 (3)		1.5		
		10 -50 µg/L		4.8 (3)		2.2		
		50.1-100 µg/L		66.4 (2)		4.8		
		>100 µg/L		134.1 (7)				
			Transitional-cell carcinoma mortality	Incidence rate (number of cases)		Relative risk:		
		≤10 µg/L		12.5 (1)		1.9		
		10 - 50 µg/L		14.9 (1)		8.2		
		50.1-100 µg/L		66.4 (2)		15.3		
		>100 µg/L		114.9 (6)				
Lewis et al. 1999a	Utah	Cumulative arsenic exposure during residence in study towns:	Bladder and other urinary organ cancers	Male	Female	Male	Female	Retrospective cohort mortality study of 4,058 residents of Millard County, Utah, born between 1900 and 1945; vital status followed through 1996 SMRs used Utah state rates as the referent
				3	2			
		<1,000 ppb-yr				0.36	1.18	
		1,000-4,999 ppb-yr				—	—	
		>5,000 ppb-yr				0.95	1.10	
		All				0.42 (0.08-1.22)	0.81 (0.10-2.93)	

	Kidney cancer	Male	Female	Male	Female
< 1,000 ppb-yr		9	4	2.51	2.36
1,000-4,999 ppb-yr				1.13	1.32
> 5,000 ppb-yr				1.43	1.13
All				1.75 (0.80-3.32)	1.60 (0.44-4.11)

Cohort Studies Reviewed in *Arsenic in Drinking Water* (NRC 1999)

Chiou et al. 1995	SW Taiwan	Cumulative arsenic: Lung-cancer incidence		Relative risk:	Incidence among a cohort of 2,556 subjects (263 blackfoot-disease patients and 2,293 healthy individuals) followed for 7 years
		<0.1 mg/L × yr	3	1.0	
		0.1-19.9 mg/L × yr	7	3.1 (0.8-12.2)	
		≥20 mg/L × yr	7	4.7 (1.2-18.9)	
		Average arsenic:			
		<0.05 mg/L	5	1.0	
		0.05-0.70 mg/L	7	2.1 (0.7-6.8)	
		≥0.71 mg/L	7	2.7 (0.7-10.2)	
		Cumulative arsenic: Bladder cancer incidence		Relative risk:	
		<0.1 mg/L × yr	4	1.0	
		0.1-19.9 mg/L × yr	7	2.1 (0.6-7.2)	
		≥20 mg/L × yr	9	5.1 (1.5-17.3)	
		Average arsenic:			
		<0.05 mg/L	6	1.0	
		0.05-0.70 mg/L	7	1.8 (0.6-5.3)	
		≥0.71 mg/L	7	3.3 (1.0-11.1)	

Continued

TABLE 2-1 *Continued*

Study	Location	Exposure[a]	End Point	Cases, No.	Study Outcome	Comments
Tsuda et al. 1995	Niigata Prefecture, Japan	<0.05 µg/L 0.05-0.99 µg/L ≥1.0 µg/L	Lung cancer mortality	0 1 8	SMR: 2.3 (0.1-13.4) 2.3 (0.1-13.4) 15.7 (7.4-31.0)	Exposure among 113 persons who drank from industrially contaminated wells for approximately 5 yr (1955-1959), then followed for 33 yr; referent is mortality in Niigata Prefecture, 1960 to 1989
		<0.05 µg/L 0.05-0.99 µg/L ≥1.0 µg/L	Bladder cancer mortality	0 0 3	SMR: 0.0 (0-12.5) 0.0 (0-47.1) 31.2 (8.6-91.8)	
Cuzick et al. 1992	United Kingdom	Total arsenic exposure: 224 mg 504 mg 963 mg 1,901 mg 3,324 mg	Bladder cancer mortality	1 1 1 1 1	SMR: 1.20 (0.04-7.0) 5.00 (2.0-15) (exposure ≥500 mg)	478 patients treated with Fowler's solution (potassium arsenite)
			Lung cancer mortality	14	SMR: 1.00 (0.5-1.7)	

Case-Control Studies Published Since *Arsenic in Drinking Water* **(NRC 1999)**

Study	Location	Exposure[a]	End Point	Cases, No.	Study Outcome	Comments
Karagas et al. 2001	New Hampshire	Toenail arsenic concentrations:	Basal-cell carcinoma incidence	587	Odds ratio: close to 1 below 97th percentile; 1.44 (0.74-2.81) above the 97th percentile toenail arsenic concentration (corresponds to > 30 g/L)	Case-control study of skin cancer; standard logistic regression used to calculate odds ratios

Reference / Country	Endpoint / Exposure	N	Results	Comments
	Squamous-cell carcinoma incidence	284	Odds ratio: close to 1 below 97th percentile 2.07 (0.92-4.66) above the 97th percentile toenail arsenic concentration (corresponds to > 30 g/L)	
Ferreccio et al. 2000 Northern Chile	Lung cancer incidence Individual 40+ year average arsenic concentration from public water supply records: 0-10 µg/L 10-29 µg/L 30-49 µg/L 50-199 µg/L 200-400 µg/L	151 cases 419 controls	Age- and sex-adjusted odds ratios: 1 1.6 (0.5 - 5.3) 3.9 (1.2 - 12.3) 5.2 (2.3 - 11.7) 8.9 (4.0 - 19.6)	Case-control study of incident lung cancer in eight public hospitals of 151 cases and 419 controls. Odds ratios calculated using subjects with average exposures of 0-10 µg/L as referent
Kurttio et al. 1999 Finland	Bladder cancer cases Average arsenic exposure: <0.1 µg/L 0.1-0.5 µg/L ≥0.5 µg/L	61 total 23 19 19	Cumulative dose for data presented (i.e., shorter latency period) calculated from beginning of the use of well water until 2 yr before the cancer diagnosis) 1.53 (0.75-3.09) 2.44 (1.11-5.37)	Population-based case-cohort study of individuals who drank water from their own drilled wells between 1967 and 1980; cases were diagnosed between 1981 and 1995; reference group of 275 age- and sex-matched persons from the same population; relative risks were calculated using subjects with exposures <0.1 µg/L as a referent population.
	Kidney cancer cases <0.1 µg/L 0.1-0.5 µg/L ≥0.5 µg/L	49 total 23 12 14	0.78 (0.37-1.66) 1.49 (0.67-3.31)	

Continued

TABLE 2-1 *Continued*

Study	Location	Exposure[a]	End Point	Cases, No.	Study Outcome	Comments
Case-Control Studies Reviewed in *Arsenic in Drinking Water* (NRC 1999)						
Bates et al. 1995	Utah	Cumulative lifetime exposure:	Bladder-cancer incidence	117 cases 266 population-based controls	Odds ratios (relative to a lifetime exposure of <19 mg):	Exposures estimated by linking residential-history information with water-sample data from public water supplies; 81 out of 88 towns <10 μg/L; 1 town >50 μg/L
		19 mg to <33 mg			1.56 (0.8-3.2)	
		33 mg to <53 mg			0.95 (0.4-2.0)	
		>53 mg			1.41 (0.7-2.9)	

[a] Units used to describe exposures in various studies, as reported by the study authors.
[b] Transitional-cell carcinoma only.
Abbreviations: β(SE), regression coefficient (standard error); SMR, standardized mortality ratio; COPD, chronic obstructive pulmonary disease.

subject, and contact between the chemical and the subject (i.e., ingestion of water or food, dermal contact, or inhalation). Dose refers to the amount of chemical transferred to the exposed subject. Dose should be described in terms of its relationship to the exposed subject, whether as the available, administered, absorbed, or active or biologically effective dose.

Measurement of the dose (as opposed to simply the fact of exposure) does not just provide the opportunity for a more detailed and informative description of the relationship between exposure and a disease; it is also important in inferring the absence or presence of a cause-and-effect relationship. One limitation of most epidemiological studies of arsenic is that dose is estimated from measuring the concentration of arsenic in drinking water and assuming an exposure based on drinking-water rates. Details of the exposure assessment conducted in each study are discussed below. The variability and uncertainty associated with the exposure assessment and their impact on a human-health risk assessment for arsenic are discussed in Chapters 4 and 5.

Chiou et al. 2001 Study

Chiou et al. (2001) conducted a prospective cohort study of 8,102 persons in the Lanyang basin of northeastern Taiwan, a region where arsenic-contaminated shallow wells were used for drinking water from the late-1940s through the mid-1990s. All cohort members had used shallow private wells for their primary drinking-water supply. Subjects were interviewed at home between October 1991 and September 1994. There were 4,586 homes represented in the study and 3,901 well-water samples taken and analyzed for arsenic with the use of hydride generation with flame atomic absorption spectrometry. Samples were not taken from 1,136 homes because their wells no longer existed. No biomarkers of arsenic exposure were used. The interview included information on history of well-water consumption, residential history, sociodemographic characteristics, cigarette smoking, personal and family medical history, occupation, and other potential risk factors. Subjects were followed for cancer incidence from the time of enrollment (1991 to 1994) through December 31, 1996, with the use of multiple resources. Detailed analyses of total urinary cancer (includes kidney, bladder, and urethral cancer) and specifically of the most common cell type of urinary cancer, transitional cell carcinoma (TCC), are presented. Nine subjects were diagnosed with bladder cancer, eight with kidney cancer, and one with both. Among those 18 subjects with urinary-tract cancer, 17 had pathological confirmation, and 11 were diagnosed with TCC. The most useful analyses were internal comparisons. Age- and sex-adjusted rates of urinary-tract cancer and TCC were first calculated within each exposure stratum. Incidences of urinary-tract cancer for subjects who drank well water with arsenic concentrations of 10.0 or less, 10.1-50.0, 50.1-100.0, and greater than 100.0 µg/L were 37.6 (three subjects), 44.8 (three subjects), 66.4 (two subjects), and 134.1 (seven subjects) per 100,000, respectively. The corresponding incidences for TCC were 12.5 (one subject), 14.9 (one subject), 66.4 (two subjects), and 114.9 (six subjects) per 100,000, respectively. When evaluated by duration of exposure to well water (<20.0, 20.1-39.9, and 40.0 years), rates for urinary cancer were 61.1 (one subject), 46.7 (four subjects), and 77.3 (10 subjects) per 100,000 persons, and rates for TCC were 0, 46.7 (four subjects), and 46.4 (six subjects) per 100,000, respectively. Cox's proportional hazards regression analysis (adjusted for sex, age, and cigarette smoking) was conducted to estimate relative risks and 95% confidence intervals by concentration and duration of arsenic exposure.

The multivariate relative risks for urinary cancer by concentration of arsenic in drinking water, using as a nonexposed referent group subjects with exposures at 10 µg/L or less, were 1.6 (95% CI = 0.3-8.0) for arsenic concentrations at 10.1-50.0 µg/L, 2.3 (95% CI = 0.4-14.4) at 50.1-100.0 µg/L, and 4.9 (CI = 1.2-15.3) at greater than 100.0 µg/L. Corresponding relative risks for TCC were 1.9 (0.1-32.2), 8.1 (0.7-98.2), and 15.1 (1.7-138.5). In a time-window analysis, the increase in arsenic-induced TCC was more prominent for subjects who had drunk well water for more than 40 years, implying the possibility of a long latency between arsenic exposure and the occurrence of TCC and/or the importance of cumulative dose.

A strength of this study is that multiple follow-up resources were used to detect newly diagnosed (incident) cancer cases, ensuring that most newly diagnosed cancers were correctly identified. Other strengths are that residential and water-use histories were obtained from most cohort members and that individual information on other risk factors, such as cigarette smoking, was available. The authors used appropriate analytical techniques to estimate relative risks and 95% confidence intervals. The approach to exposure estimation has both strengths and weaknesses. The study was conducted in an area of Taiwan with a residentially stable population. Under the assumption that arsenic concentrations from household wells had been relatively constant over time, past individual exposures could be estimated by measuring arsenic in the current household well water of most cohort members. It should be noted that few data directly address the issue of arsenic-concentration consistency over time, and the question of historical consistency of arsenic concentrations in various types of groundwater deserves exploration. Consistency is of particular concern with shallow groundwaters, which might be subject to greater fluctuation than water from deeper wells. There were 4,074 study subjects (50.3%) who had drunk well water for more than 40 years, and over 70% of subjects greater than 50 years of age had drunk well water for more than 30 years. The average duration of drinking well water was 40.7 years. For 2,119 subjects (26.2%), more than one well had been used in their houses, and their past arsenic exposure data were derived from their current well data (Cantor 2001). Those 2,119 subjects had used their current well for approximately 60% of their life spans (standard deviation = 21 years), and most had used their current well for at least 10 years (Chen and Chiou 2001). Although exposure assessment was individualized, for all subjects only a single well-water measurement obtained at one point in time was used to characterize long-term arsenic exposure.

Foremost among the limitations of this study is the short duration of follow-up, which limited the number of person-years of observation; hence, only a few cases of incident urinary-tract cancer were recorded: a total of 18 cases (15 with exposure information), of which 11 were TCC (10 with exposure information). In calculating relative risks, the small number of cases resulted in very wide CIs. For example, the adjusted relative risk of TCC in the highest exposure category (>100.0 μg/L) was 15.1, with a 95% CI of 1.7-138.5. Thus, the study by Chiou et al. (2001) is limited by its relatively small numbers, and risk estimates based on these data might be too imprecise for use in a quantitative risk assessment. However, the data can serve as supplementary information, along with data from other selected studies. Since the cohort continues to be studied by the investigators, it is likely to yield more stable risk estimates in the future as more person-years of follow-up are accrued.

Ferreccio et al. 2000 Study

In a region of northern Chile with a history of increased concentrations of arsenic in drinking water, Ferreccio et al. (2000) conducted a case-control study of incident lung cancer. Eligible cases included all lung cancer cases diagnosed in the eight public hospitals in regions I, II, and III from November 1994 through July 1996. There were 217 eligible subjects identified, and 151 (70%) participated. Nonparticipation was largely due to inability to locate the subject. Controls were selected from the hospitals where the lung cancer cases were diagnosed. Two control series were selected: cancers other than lung cancer and noncancer controls. Another set of similarly identified controls was selected for a parallel study of incident bladder cancer. Cases and controls were interviewed regarding residential history, socioeconomic status, occupational history (to ascertain employment in copper smelting), and smoking.

An unusual characteristic of this study is the method used for control selection. Because control selection is a central element of the study design and might influence interpretation, it is worthwhile to describe the selection procedure. If it is assumed that most cases of lung cancer that occurred in the population were included in the study, an appropriate control group would be healthy persons selected randomly from the population at large, matched on age and sex. Instead, the authors selected hospitalized non-lung-cancer patients as controls. These patients could be more readily identified and interviewed. This technique has frequently been used in cancer epidemiology and can be an acceptable approach. In an effort to avoid overmatching by geographic area, and thus relative concentrations of arsenic in drinking water, the authors sought to select controls from the study hospitals in a manner that would reflect the distribution of arsenic exposure in the overall population from which the cases arose. To do that, the probability of selecting a control from each of the eight hospitals was based on the relative frequency of admission to that institution in 1991. Controls were also matched on sex and age within 4 years of the index case. Two hospital-based controls were selected for each case using this method: a patient diagnosed with a cancer that has not been related to arsenic exposure and a noncancer patient.

Historical exposure to arsenic in drinking water for each respondent was estimated by linking residential history information with a database of information on arsenic concentrations in public-water supplies collected for the years 1950 through 1994. Arsenic concentrations in the years prior to 1950 were based on concentrations in the 1950s. Average arsenic concentration in the place of residence was assigned to each subject on a year-by-year basis for the period of 1930-1994. Population coverage of public-water systems in the main cities in regions I and II was over 90% and was between 80% and 90% in the major cities of region III. The coverage in smaller cities varied between 64% and 91%.

The authors conducted a series of validation checks for the controls. Control- group distribution among hospitals was compared with the target distribution based on admissions in 1991. Major discrepancies were found, the main differences being from the main hospitals of Arica and Antofagasta, the two largest cities in the study area. There was a deficit in the proportion of observed-to-expected controls from the Arica hospital (0.6) and an excess from the Antofagasta hospital (1.2). In another comparison, the authors used the distribution of arsenic concentration in the period of 1958-1970 based on population figures from the 1992 census. Population numbers in 1992 were used to estimate an expected water-arsenic distribution for 419 randomly selected controls. This distribution was then compared with the actual distribution of the selected controls, and the ratio of the selected numbers of controls to that expected was calculated. The baseline exposure stratum (0-49 μg/L) showed a selected-to-expected ratio of 0.8. However, the high-exposure category of 400 μg/L and above (selected-to-expected = 1.4) was overrepresented, owing to overselection of controls in Antofagasta, which had unusually high concentrations of arsenic in its drinking water during 1958-1970. Selected-to-expected ratios for intermediate groups were 1.3 for 50-99 μg/L and 0.5 for 100-399 μg/L. The expected impact of the overselection in the highest-exposure category is to diminish estimates of risk, and underselection in the lower stratum is to enhance risk estimates. As the authors noted, this assessment is indirect, because in the data analysis the actual historical residential location of cases and controls rather than their location at the time of the study was used to determine arsenic exposure levels. However, the direction of bias is probably as indicated.

Odds ratios were used to estimate relative risk of exposure to various concentrations of arsenic in drinking water relative to a referent concentration of 0-10 μg/L. Odds ratios were calculated using unconditional logistic regression, adjusted for age, socioeconomic status, smoking, and working in a copper smelter. In separate analyses, arsenic in drinking water was expressed as average yearly concentration during the peak years of exposure 1958-1970 or as average yearly concentration in the period of 1930-1994. In Antofagasta, there was a peak in exposure during 1958-1970, when arsenic concentrations averaged 860 μg/L. When 1958-1970 average arsenic concentrations were used as the estimate of exposure, the odds ratios were 5.7 (95% CI = 1.9-16.9) for the 400-699 μg/L exposure stratum and 7.1 (95% CI = 3.1-12.8) for 700-999 μg/L, relative to 0-10 μg/L. Results from the analysis based on average exposures during 1930-1994 and using all controls (419) showed an increase in the odds ratio with arsenic concentration, reaching an odds ratio of 8.9 (95% CI = 4.0-19.6) for the highest exposure group (average exposure 200-400 μg/L). Results from calculations using cancer controls (n = 167), noncancer controls (n = 252), or controls frequency-matched to lung cancer cases (n

= 237) were of similar magnitude. Risk estimates based on long-term average exposure were higher per unit exposure than those that used the peak period of exposure (the mid-point of which is about 30 years before diagnosis of study cases). It is not clear at this time whether it is appropriate to give more weight to the risks based on 40+ years of exposure or to risks based on the 12-year peak exposure period about 30 years prior to diagnosis. When the study population was stratified by smoking status, there was suggestive evidence of a synergistic interaction between smoking and exposure to arsenic in drinking water. Relative to nonsmokers with average arsenic exposures of 49 µg/L or less (1930-1994), nonsmokers with average exposures of 200 µg/L or more had an odds ratio of 8.0 (95% CI = 1.7-52.3), whereas smokers with average exposures of 200 µg/L or more had an odds ratio of 32.0 (95% CI = 7.22-198.0).

Strengths of this study include an acceptable response rate, unbiased ascertainment of exposure, individual estimates of exposure, exposure coverage of most of the life span for most study subjects, incorporation of individual data on other potentially confounding risk factors for lung cancer, appropriate analyses of study data, and an adequate study size. This is the only study available for risk assessment that has individual estimates of exposure on all subjects for more than 40 years, well beyond the minimum latency for lung cancer. The major limitation of the study, as discussed, is related to methods used for control selection. The data from this study can be used in a quantitative assessment of risk of arsenic in drinking water, along with data from other selected studies.

Lewis et al. 1999 Study

Lewis et al. (1999a) conducted a retrospective cohort mortality study of 4,058 residents of Millard County, Utah, a region where drinking water is derived from wells and where arsenic concentrations in well water range from undetectable to up to a few hundred micrograms per liter. In a letter to the editor (Lewis et al. 1999b), the purpose of the study is described as follows: "We were interested in determining whether studies of health effects related to arsenic in drinking water could be conducted in U.S. populations exposed to relatively low concentrations of arsenic." The study cohort included two groups: 2,073 persons from an earlier study (Southwick et al. 1982) and 1,985 additional individuals identified from special census records and other records maintained by the Church of Latter Day Saints (Mormons). The earliest entry in the cohort was in 1900, and the most recent was in 1945. Cohort members had lived in the towns of Delta (1,191 persons; 29.4%), Hinckley (1,192 persons; 29.4%), or smaller towns, such as Deseret, Abraham, or Oasis (1,675 persons; 41.2%). Vital status follow-up was through November 27, 1996. At the closing date, 38.2% (1,551) were alive, 54.3% had died (2,203), and 7.4% (300) were lost to follow-up. All death certificates were coded according to rubrics of the ICD-9, and SMRs were calculated by cause of death, separately for males and females in each of three exposure groups, as described below, and for all exposure groups combined. The SMR is the cause-specific ratio of observed-to-expected numbers of deaths, where the expected number is calculated for the age and sex distribution of the study population by using age- and sex-specific rates of a comparison population. In calculating SMRs, the comparison was sex-specific rates for the state of Utah. That metric was calculated by multiplying the concentration of arsenic in the drinking water in parts per billion (micrograms per liter) by the number of years of exposure to yield an exposure with units of parts per billion-years, arranged into three groups: <1,000, 1,000-4,999, and 5,000 ppb-yr. This metric represents an estimate of cumulative arsenic exposure for years of residence in study towns. Information on water intake was not available. For several causes of death, statistically significant increases in SMRs occurred in a generally uneven pattern based on exposure level. Among men, the causes of death included hypertensive heart disease, nephritis, nephrosis, and prostate cancer. Among women, they were hypertensive heart disease and all other heart disease (pulmonary heart disease, pericarditis, and other diseases of the pericardium).

This study has several strengths. Vital status was determined for all but a small number of cohort members (7.4%), selection into the cohort was unbiased with respect to outcome, and numbers of cohort members and deaths within the cohort were adequate to evaluate risk for major causes of death.

Several of the limitations concern the estimation of exposure and the consequences for study interpretation. Lewis et al. (1999a) present results of 151 arsenic analyses of drinking water in public and private water supplies for the study towns from a survey of drinking-water wells conducted by the Utah State Health Laboratory since 1976. These data "were used in assessing the potential exposure of cohort members to arsenic in drinking water" (Lewis et al. 1999a). Lewis et al. (1999a) present the number of samples taken in each town and the average, median, minimum, and maximum amounts of measured arsenic. The number of wells that contribute to the municipal water supply of each town is not presented, nor is it stated whether multiple measurements in each town represent repeat samples taken from one well, multiple samples from different wells, or a combination of both. For each year a cohort member lived in one of the study communities, he or she was assigned the median concentration of measured arsenic for the wells of that community. Presumably, exposure that occurred during periods when a person did not live in one of the study communities was not included in the estimate of exposure used in the analysis, creating the possibility that cumulative lifetime arsenic exposure might have been incompletely ascertained for some of the cohort.

The range of arsenic concentrations reported in Lewis et al. (1999a) for each town was quite broad. For example, arsenic concentrations measured since 1976 in Hinckley ranged from 80 to 285 µg/L; in Delta, 3.5 to 125 µg/L; and in Deseret, 30 to 620 µg/L. However, the median concentration was assigned to individual cohort members during their years of residence. Use of a median level to estimate exposure when the concentrations vary so widely necessarily resulted in misclassification of exposure for many subjects. A similar approach to estimating exposure was taken in the ecological studies from southwestern Taiwan of Chen et al. (1985) and Wu et al. (1989) that were used in the EPA risk assessment. The estimates of exposure from the Taiwanese studies might be more representative of actual exposures than the estimates from Utah. In Utah, arsenic samples came from different types of water supplies, both public and private.

The exposure metric used for the analysis (parts per billion-years) was the product of water arsenic concentration (in parts per billion) and duration of exposure (in years). One consequence of using this exposure metric, which inextricably combines intensity and temporal aspects of exposure, is that comparisons of results with findings of many other studies that used measurement of average arsenic over extended periods are not possible. Another consequence of using this metric is that persons with very different exposure histories are grouped in the same exposure stratum during data analysis. For example, a value of 1,000 ppb-yr could result from exposure to 20 µg/L for 50 years or 200 µg/L for 5 years. Such strikingly different exposure scenarios could have very different health consequences. Other studies (e.g., Chiou et al. 2001) suggest that average arsenic concentration over a long period is more strongly associated with risk than is the duration of exposure to increased concentrations, suggesting the need to consider concentration and duration of exposure independently. The duration of exposure for cohort members is not explicitly presented and might have been relatively brief for some.

In addition to limitations regarding exposure ascertainment, the subcommittee noted a few epidemiological issues. As the authors recognized in the discussion, "SMRs cannot be directly compared in an analysis that uses indirect adjustment," especially where the distributions of age among groups are not comparable. There was a significant difference in the age distribution of the three exposure groups, indicating that comparison of SMRs across exposure groups is not appropriate. The difference in age distribution is not surprising among exposure groups in which the exposure metric itself is age-dependent. Older people would be likely to have higher cumulative exposure. "Based on this, any conclusions on whether arsenic is an etiologic factor in consideration of increased or decreased SMRs among the groups is uncertain" (Lewis et al. 1999a).

Comparison rates used in the analysis of risk were for the state of Utah, and this comparison likely resulted in underestimates of risk for some causes of death and overestimates for others. The study cohort was composed of Mormons, with strict religious prohibitions against smoking and consumption of beverages containing alcohol or caffeine. In addition, the study population was largely rural. As the study authors noted, smoking rates for the state of Utah are low, around 12-13%. However, even this relatively low smoking prevalence is expected to result in rates of several cancers (lung, bladder, kidney, and pancreas) and cardiovascular diseases that are increased relative to a nonsmoking population, such as the study cohort. That factor and the rural setting of the cohort were possible contributors to the deficits observed in SMRs for urinary and pulmonary cancers.

A reanalysis of the study data for bladder and lung cancers, without a comparison population, was conducted by the EPA Office of Water (EPA 2000) using Cox proportional hazards regression. The Lewis et al. (1999a) study identified a total of 34 respiratory cancers, of which 30 were lung cancers, and a total of five bladder cancers. The risk detected in the EPA reanalysis was not statistically distinguishable either from zero or from the levels predicted by model 1 of Morales et al. (2000) and used by EPA in its arsenic risk analysis (EPA 2000). In addition to limitations in exposure assessment, it is apparent that the power of the Utah study is too low to allow a precise enough estimate for use in quantitative risk assessment. It is possible that further exploration of the exposure assessment used for this study would be fruitful. If so, then additional follow-up and reanalysis using an internal referent is to be encouraged. In summary, although this study in its current state has several strengths, several limitations preclude its use for quantitative risk assessment.

Tsai et al. 1999 Study

Sex-specific mortality due to several cancer and noncancer causes in the area of southwestern Taiwan where blackfoot-disease is endemic was evaluated by Tsai et al. (1999) for the years 1971 through 1994. Formerly, this area had high concentrations of arsenic in drinking water. Standard mortality ratios were calculated twice using two referent groups: the first referent was the mortality experience of the whole of Taiwan and the second referent was the mortality experience of the two counties of southwestern Taiwan where blackfoot disease is endemic. This study has special significance because it speaks to the issue of possible factors, such as differences in diet, cultural background, smoking, occupational exposures, access to medical care, or other differences (other than arsenic exposures) between the population of southwestern Taiwan and the remaining population of the country that might have influenced cancer mortality. In particular, it has been argued that differences in general nutritional status or selenium intake between the relatively poor farming and fishing communities of southwestern Taiwan (the locale of high arsenic exposures from water and the studies used for the EPA risk assessment) and the remainder of the country were great enough to result in increases of risk that might erroneously be attributed to drinking-water arsenic exposures. Those concerns apparently led EPA, in conducting its risk assessment, to reject the notion that the mortality experience of the whole Taiwanese population was appropriate as a low-exposure referent. Although there remains some possibility that population differences in nutrition, socioeconomic status, or other factors between southwestern Taiwan and the remainder of the country have some influence on their respective cancer rates, results from the sex- and cancer-specific calculations of Tsai et al. (1999) provide evidence that such differences are relatively unimportant. As examples, the subcommittee used the SMRs from Tsai et al. (1999) for three cancer sites of interest—lung, bladder, and kidney cancers among males and females—that were calculated using regional and national cancer rates, respectively, as the referent. Among males in the area where blackfoot-disease is endemic, there were 699 deaths due to lung cancer. Using regional rates as the referent, the lung cancer SMR for males was 3.10 (95% CI = 2.88-3.34), and using national rates, the SMR was 2.64 (95% CI = 2.45-2.84). There were 312 male deaths due to bladder cancer. Using regional rates as the referent, the SMR for bladder cancer was 8.92 (95% CI = 7.96-9.96), and using national rates, the SMR was 10.50 (95% CI = 9.37-11.73). There were 94 male deaths due to kidney cancer, the corresponding SMRs being 6.76 (95% CI = 5.46-8.27) using regional rates, and 6.80 (95% CI = 5.49-8.32) using national rates. Among females, there were 471 deaths due to lung cancer. The SMR for women was 4.13 (95% CI = 3.77-4.52) using regional rates as the referent, and the SMR was 3.50 (95% CI = 3.19-3.84) using national rates. There were 295 female deaths due to bladder cancer. The SMR for bladder cancer was 14.07 (95% CI = 12.51-15.78) using regional rates, and the SMR was 17.65 (95% CI = 5.70-19.79) using national rates. There were 128 kidney cancer deaths among women. SMRs for kidney cancer mortality were 8.89 (95% CI = 7.42-10.57) using regional rates and 10.49 (95% CI = 8.75-12.47) using national rates. As noted by Tsai et al. (1999), the regional referent population was very similar to the area where blackfoot-disease is endemic in lifestyle factors, and thus the similarity of SMRs analyzed with either the regional control population or the national population indicates that significant confounding was unlikely.

Finally, it is interesting to note that the study by Tsai et al. (1999) determined that the area of southwestern Taiwan where arsenic is endemic experienced significant increases in mortality from many other cancers besides those commonly linked to arsenic. It is possible that the relatively high arsenic exposure of the study population and the large size of the study in terms of person-years of observation might have enabled these statistically significant increases to emerge.

Kurttio et al. 1999 Study

Water samples from drilled wells in some regions of Finland have been reported to have high concentrations of inorganic arsenic. To test whether drinking water from such sources was associated with risk of bladder or kidney cancer, Kurttio et al. (1999) conducted a population-based case-cohort study in Finland of 61 bladder cancer and 49 kidney cancer cases. Cases, diagnosed in the period of 1981-1995, were selected from among Finnish residents of places not served by municipal water supplies who had used drinking water from their own drilled wells between 1967 and 1980 and for whom a water sample was analyzed for arsenic concentrations. A reference group of 275 persons, matched by age and sex to the combined case series, was selected from the same population. Contrary to expectation, arsenic concentrations in the cancer group and the control group were relatively low. Among the 61 bladder cancer cases, 42 drank water from wells with arsenic below 0.5 µg/L. The 95th percentile of the distribution of arsenic in well water among bladder cancer cases was 3.0 µg/L, and among the referent cohort, 4.5 µg/L. In one analysis, the study population was stratified by period of exposure—one group with exposure in the third to ninth calendar year before cancer diagnosis (shorter latency) and another group with exposure in the tenth calendar year and earlier before diagnosis (longer latency). Among persons in the shorter latency group, a relative risk of 2.44 (95% CI = 1.11-5.37, 19 cases) was observed for an average arsenic exposure of 0.5 µg/L or more, relative to persons with average arsenic exposure of less than 0.1 µg/L. Relative risks for the comparable exposure groups among persons with longer latency were not significantly above one. Increased risk with arsenic exposure was confined to persons who smoked during the 1970s. Among the shorter latency group, smokers with average arsenic concentrations of 0.5 µg/L or above had a relative risk of 10.3 (95% CI = 1.16-92.6); the relative risk among nonsmokers in this arsenic exposure group was 0.87 (95% CI = 0.25-3.02). Both calculations were relative to people (smokers or nonsmokers, respectively) with arsenic concentrations of less than 0.1 µg/L.

The finding of increased bladder cancer risk in this study is striking, as the risk is far above that expected at relatively low concentrations of arsenic exposure. Given the small number of bladder cancer cases in the study (61) and the likelihood that much more arsenic would come from food than water in this low-exposure population, that finding might be the result of a chance observation or an unmeasured bias.

Karagas et al. 2001 Study

Using levels of arsenic in toenails as a biomarker of internal exposure, Karagas et al. (2001) conducted a case-control study of 587 basal-cell carcinoma cases (BCC) and 284 squamous-cell carcinoma cases (SCC) in New Hampshire, where private wells are used by about 40% of the population as the primary source of drinking water. Although the arsenic concentration in a majority of New Hampshire wells is below 1.0 µg/L, the concentration in many underground water sources is higher, with concentrations above 50 µg/L in some instances. Earlier, Karagas et al. (2000) had demonstrated a correlation (r) between arsenic in toenails and arsenic in water in persons exposed to arsenic at 1.0 µg/L or more of water (r = 0.65, p < 0.001). In the case-control study, controls were randomly selected from the general population of New

Hampshire, matching on age and sex to the overall distribution of the two case series. Cases and controls were interviewed at home starting in January 1994. The questionnaire included questions on type of water supply and the number of years of use. Toenail clippings were collected at the time of the interview and analyzed using instrumental neutron activation analysis for arsenic concentration. Standard logistic regression methods were used to calculate odds ratios for BCC and SCC relative to toenail arsenic concentration, as stratified into six strata according to percentile of the overall distribution among controls. The odds ratios for SCC and BCC were close to unity in all but the highest category. Among subjects with toenail arsenic concentrations above the 97th percentile, the adjusted odds ratios were 2.07 (95% CI = 0.92-4.66) for SCC and 1.44 (95% CI = 0.74-2.81) for BCC, compared with those with concentrations at or below the median.

The concentration of drinking-water arsenic that corresponds to the 97th percentile lower cutoff level for the highest stratum of toenail arsenic in this study (0.345 µg/g), where increased odds ratios were observed, is approximately 30-50 µg/L (estimated by reference to the earlier publication of Karagas et al. 2000), showing a correlation between water and toenail arsenic concentrations in controls in that study. The highest level of toenail arsenic among controls was 0.81 µg/g, and given the overall log-normal distribution of toenail arsenic in this population, the median level of toenail arsenic in this stratum was likely to be about 0.450 µg/g, corresponding to approximately 100-150 µg/L of drinking water.

There are several uncertainties regarding the exposure assessment in this study. Arsenic bound to the toenail matrix can result from external exposure as well as intake of inorganic arsenic in food and water. These and other factors result in significant variability in toenail arsenic levels among individuals with similar exposure to arsenic in drinking water. Little is known about the toxicokinetics of arsenic uptake into toenails. In the setting of this study, that lack of knowledge leads to uncertainty about the concentration of arsenic in drinking-water exposures corresponding to toenail arsenic concentrations at which increased risk was observed. Arsenic in toenail clippings typically integrates exposures over a few week period about a year prior to sample collection. The latency for skin cancer is not well defined, but appears to be more than a decade. Thus, the time period of exposures reflected in toenail arsenic does not correspond to the critical exposure period for skin cancer. Over 50% of the study population had used the same water supply for over 15 years. Relative-risk estimates did not vary significantly according to length of use by subjects, further complicating interpretation of study results. An additional limitation of this study is the relatively low control participation rate, 50%, which might have resulted in a biased sample from the general population.

Tucker et al. 2001 Study

Tucker et al. (2001) analyzed data describing prevalent skin cancer and other skin disorders (hyperkeratoses and dyspigmentation) from a cross-sectional study in a region of Inner Mongolia with increased concentrations of arsenic in drinking water. A total of 3,179 persons in three villages were examined in 1992, and well-water-use histories for these individuals were gathered. Water samples were collected from 184 of the 187 local wells in these villages and analyzed for arsenic content. The median age of participants was 29 years, and the average well-use history was 25 years. In the study population, 79.6% were less than 50 years of age. Arsenic in drinking water ranged from below detection (10 µg/L) to 2,000 µg/L. Skin cancer was observed in eight subjects. In addition, 172 subjects had keratosis, 121 had dyspigmentation, and 94 subjects were diagnosed with both types of lesions. Among the skin cancer cases, all eight had both keratosis and dyspigmentation. Two exposure metrics were used in the data analysis: peak arsenic concentration (PAC), defined as the highest well-water arsenic concentration reported by an individual, and cumulative-arsenic dosage (CAD), estimated by multiplying the arsenic concentration by the duration of well use for each well reported to have been used by each subject and then adding the products together to get cumulative exposure. Thirty-five percent of the study population had a PAC exposure of less than 50 µg/L, and 86% had a PAC exposure of less than 150 µg/L. Several statistical models (frequency weighted, simple linear regression, hockey stick, and maximum likely estimate) were used to analyze data for each of the effects noted. Two measures of exposure were used with each model.

Dose-response relationships were found for skin cancer and the other health end points, using both metrics of exposure to arsenic in drinking water. Several of the statistical models used in the analysis of these data used an a priori assumption of a threshold exposure concentration of arsenic in drinking water below which skin cancer (and other dermatological effects) would not occur. The data appear to be adequately described by such statistical models; however, they are also well-described by a nonthreshold linear model. That is not surprising in view of the small number (eight) of skin-cancer cases observed. The authors do not discuss whether differences in results among the various models used to describe the data were statistically significant.

There are limitations in the two exposure metrics (CAD and PAC) that were used in this study. Limitations in using CAD as a measure of exposure are described above in the discussion of the Lewis et al. (1999a) study. The other exposure metric, PAC, was used without reference to the time of peak exposures. That lack of information is problematic, because many of the study subjects might have used more than one well during relevant exposure periods, and the peak exposures might have occurred many years before this cross-sectional study and for as little as 1 year.

LATENCY PERIOD

For cancers, there is generally a period of time, which may be as long as several decades, between a critical molecular interaction of a carcinogen within a single cell and the first appearance of a malignant cell. The length of this period will vary based on the nature of the cancer and whether the molecular interaction occurs early or late in the chain of events leading to the cancer. That period between the critical exposure to the carcinogenic agent and the occurrence of the cancer is referred to as the latency period. Latency is usually estimated by determining the period between the time of first exposure to the carcinogen and the clinical detection of the cancer, and studies have been conducted that provide information on the latency period of arsenic-induced skin cancer and internal cancers.

Because skin cancers are visible, the time of their appearance can often be dated with some accuracy, which makes the estimation of the latency period easier for these cancers. A number of early studies that examined latency for arsenic-induced skin cancer were on patients who had ingested arsenical medications; therefore, those studies have accurate information on the time of exposure (see Table 2-2).

Neubauer (1947) summarized the published data for 143 cases of skin cancer (epitheliomas: basal- and squamous-cell carcinomas) attributed to ingestion of medicine containing arsenic. Latency periods between starting the arsenical medication and detection of cancer ranged from 3 to 40 years, with a mean of 18.1 years. Sommers and McManus (1953) presented data on 22 patients who had an average latency period of 24 years (range of 13-50 years) between first administration of arsenical medication or occupational

TABLE 2-2 Estimates of Latency in Studies of Skin Cancer Associated with Arsenic Ingestion

Study	Study Type	No. of Subjects	Place	Estimated Latency (years)
Neubauer 1947	Case-series	143	Various	Mean, 18; range, 3-40
Sommers and McManus 1953	Case-series	22	U.S.	Mean, 24; range, 13-50
Fierz 1965	Case-series	21	Germany	Mean: 14
Wong et al. 1998	Case-series	17	Singapore	BD: mean, 39; range, 29-50 SCC: mean, 41; range, 32-47
Smith et al. 1998	Ecological	27 deaths	Chile	Range, about 20-35
Boonchai et al. 2000	Case-series	34	Australia	Mean, 20; range, 6-39

Abbreviations: BD, Bowen's disease; SCC, squamous-cell carcinoma.

exposure to arsenic-containing pesticide sprays and appearance of skin cancers. Fierz (1965) estimated a mean latency of 14 years for 21 patients with skin cancers. That estimate, however, is probably subject to bias, because the cases were self-selected from a much larger cohort of patients (1,450) who had received arsenical medication. In addition, the follow-up time was only a maximum of 26 years, which probably resulted in an underestimate of the mean latency.

More recently, Wong et al. (1998) studied 17 patients in Singapore with skin lesions who had taken Chinese medicines containing arsenic and estimated a mean latency period for Bowen's disease of 39 years (range of 29-50 years), and a mean for squamous-cell carcinoma of 41 years (range of 32-47 years in seven patients) (Wong et al. 1998). Thirty-four self-referred Australian patients who had basal-cell carcinoma and had taken an arsenical asthma medication produced a mean latency estimate of 20 years (range of 6-39 years) (Boonchai et al. 2000). Most of those studies, however, are case series and are limited by their small sample size, the possibility of underestimating latency because of lack of follow-up until the deaths of the entire case-series patients, and possible confounding from other causes of skin cancer. Smith et al. (1998) studied cancer rates in a region of Chile, which between 1958 and 1970 had a water supply with a high concentration of arsenic (above 800 µg/L). The SMRs for skin cancer for the period of 1989-1993 were 7.7 and 3.2 for males and females, respectively. Because those SMRs are for only one time period, they are difficult to interpret, but they would be consistent with a latency of around 20-35 years. However, because the SMRs were already high in the 1989-1993 time window, it is likely that mortality was already elevated before the observation period. In addition, many years usually elapse between the clinical diagnosis of skin cancer and mortality from skin cancer, implying a shorter latency for occurrence of those cancers.

In the case of internal cancers, the data are from epidemiological studies; however, few of those studies address latency (see Table 2-3).

Chen et al. (1986) conducted a case-control study of cancers in a high-arsenic area of Taiwan. They calculated age- and sex-adjusted odds ratios for cancers in individuals who consumed artesian well water contaminated with arsenic for more than 40 years. The odds ratios at earlier time periods were lower than those for later time periods. Consistent with those results, in a

TABLE 2-3 Estimates of Latency in Studies of Internal Cancers Associated with Arsenic Ingestion

Study	Study type	Cancer Type	Place	Estimated Latency (years)
Chen et al. 1986	Case-control	Bladder, lung, liver	Taiwan	>40
Cuzick et al. 1982, 1992	Cohort	Bladder	England	Mean, 7; range, 6-8 (n=3)[a] >20 (n=2)[a]
Tsuda et al. 1995	Cohort	Lung (n=9) Bladder (n=3)	Japan	Mean, 27; range, 11-35[a] Mean, 32; range, 25-37[a]
Bates et al. 1995	Case-control	Bladder	Utah, U.S.A.	Maximum; 30-39 (smokers only, based on time-window analyses)
Smith et al. 1998	Ecological	Lung, bladder, kidney	Chile	Range; about 20-35[a]
Chiou et al. 2001	Cohort	Bladder	Taiwan	>40

[a] Death as the latency end point.

recent cohort study in the same area, Chiou et al. (2001) found a higher bladder cancer risk for individuals consuming well water for 40 or more years 224-3324 mg). The latency period was between 6 and 8 years for three of the deaths, and more than 20 years for the other two deaths. Those data suggest the possibility that arsenic might have both early- and late-stage effects on the carcinogenic process, although the number of cases was small and information was lacking on the smoking histories of the individuals.

In Japan, Tsuda et al. (1995) followed a cohort of 131 people who had been exposed to well water containing high concentrations of inorganic arsenic between 1955 and 1959 and followed until 1992. In the highest exposure group ($\geq$1.0 mg/L), the SMR for bladder cancer was 31.2 (95% CI = 8.6-92) and for lung cancer 15.7 (95% CI = 7.4-31). For eight lung cancer cases, the mean latency was 27 years (range of 11-35 years); for three bladder cancer cases, the mean latency was 32 years (range of 25-37 years). Because death is the end point used to calculate the latency period, the latency would tend to be overestimated in this study.

Using a time-window analysis of bladder cancer case-control data and arsenic exposure from the state of Utah, Bates et al. (1995) found that in smokers the highest odds ratios occurred 30-39 years after exposure. A similar result was not seen for nonsmokers. However, some doubt is cast on this result because of the relatively low arsenic exposures involved.

As described above for skin cancer, Smith et al. (1998) studied cancer rates during 1989-1993 in a region of Chile. The SMRs for male bladder and lung cancers were 6.0 (95% CI = 4.8-7.4) and 3.8 (95% Cl = 3.5-4.1), respectively; the corresponding SMRs for females were 8.2 (95% Cl = 6.3-10.5) and 3.1 (95% Cl = 2.7-3.7). As with skin cancer, those data would be consistent with a latency period of around 20-35 years. As noted above, because the SMRs were already elevated in the 1989-1993 period of observation, it is likely that the increase in mortality started several years earlier. Chiou et al. (2001) studied a cohort of Taiwanese people who were exposed to arsenic and developed bladder cancer. The latency period for this cancer was estimated to be more than 40 years.

Although the data are sparse, they are generally consistent in suggesting that both skin and internal cancers associated with arsenic ingestion have, on average, substantial latency periods, frequently in excess of 20 years from the beginning of exposure. The existing data are insufficiently detailed for a more precise specification of the distribution of latencies. There has been little investigation of the influence of exposure level, duration of exposure, or age at exposure on the latency period for arsenic-induced cancers. Data for other carcinogens, however, suggest that dose might not always affect latency period, even though it affects cancer incidence (Armenian 1987).

ESSENTIALITY

The question of the essentiality of arsenic was reviewed in the 1999 NRC report. At that time, it was concluded that arsenic had not been investigated for essentiality in humans. There have been no new studies since that report identifying a potential need for arsenic in human nutrition. A recent review of dietary reference intakes noted that "[b]ecause of the lack of human data to identify a biological role of arsenic in humans, neither an estimated average requirement (EAR), Recommended Dietary Allowances, nor Adequate Intake levels were established" (IOM 2001). The IOM report made several recommendations, including that "the role of arsenic in methyl metabolism and genetic expression requires further study. Necessary for future studies with humans is the identification of a reliable indicator of arsenic status" (IOM 2001).

SUMMARY AND CONCLUSIONS

- For many chemicals, health effects must be determined on the basis of animal studies or occupational exposures. In contrast, a large number of general- population epidemiological studies have investigated the noncancer and cancer effects of chronic exposure to arsenic in drinking water.
- Since the 1999 NRC report on arsenic in drinking water, additional evidence has emerged linking arsenic consumption in drinking water with two noncancer health conditions that are a major source of morbidity and mortality: hypertension and diabetes mellitus. The recent prevalence study by Rahman et al. (1999) in Bangladesh found a dose-dependent increase in the risk of hypertension that was largely concordant with a prior prevalence study conducted by Chen et al. (1995) in the area of southwestern Taiwan where arsenic is endemic. A prospective cohort study of noninsulin-dependent diabetes mellitus (Tseng et al. 2000) in the arsenic endemic area of southwestern Taiwan found a positive association with cumulative arsenic exposure that was generally consistent with the prevalence study by Rahman et al. (1998) in Bangladesh. Although additional studies are needed to assess the dose-response relationship for these end points, it is notable that the arsenic expo-

sure associated with these substantial noncancer risks in Taiwan and Bangladesh were within 1 to 2 orders of magnitude of concentrations that are of current regulatory concern (e.g., 20 µg/L or below).

• Few studies of the effects of arsenic on reproduction and development had been published at the time of the 1999 NRC report. Since that time, a small number of studies have investigated the relationship between arsenic exposure in humans and adverse reproductive effects, including studies of populations in Chile and Bangladesh exposed to increased concentrations of arsenic in drinking water. There is some evidence from those studies that arsenic increases infant mortality, spontaneous abortions, stillbirths, and preterm births. However, the evidence is not conclusive, because the studies suffer from such limitations as a lack of information on lifestyle and other exposures that could affect reproductive outcomes. Nonetheless, the number of studies investigating whether arsenic might have adverse effects on reproduction and development is increasing, and there is suggestive evidence of effects on several outcomes.

• Findings from a large prevalence study in West Bengal, a small prevalence study in Bangladesh, and an ecological mortality study in the area of southwestern Taiwan where arsenic is endemic add to previous suggestions that arsenic ingestion might be associated with noncancer respiratory effects. The pathology that characterizes these effects has not been defined.

• Recent studies from West Bengal, Bangladesh, and Inner Mongolia have examined dose-response relationships between ingestion of arsenic in drinking water and skin lesions. Those studies have reported the presence of arsenic- related skin lesions in some study subjects consuming drinking water with arsenic concentrations of less than 100 µg/L; however, the findings might be subject to uncertainties and bias with respect to classification of the exposure and the end point.

• Two recent small investigations of neuro-cognitive function in young schoolchildren suggest that arsenic exposure might be associated with an adverse effect, but uncertainties regarding the extent of the subjects' exposure to arsenic and to other potential confounders, such as lead, limit study interpretation.

• Since completion of the previous evaluation of arsenic by the NRC (1999), several human-population-based studies have been completed. These studies confirm and extend the observations that were available to the 1999 NRC committee and further contribute to our understanding of dose-response relationships for cancers and arsenic in drinking water.

Two studies published since the 1999 NRC report (Ferreccio et al. 2000; Chiou et al. 2001) have certain strengths that go beyond the ecological studies of cancer mortality in southwestern Taiwan that served as the primary basis for the previous EPA risk assessment. These two studies evaluated risk factors among newly diagnosed cases (not deaths) of urinary cancer (Chiou et al. 2001) or lung cancer (Ferreccio et al. 2000). They incorporated individualized information on long-term arsenic exposure and health outcome. They also gathered and analyzed information on other risk factors, such as smoking habits, from individuals. That information was not available in the previous ecological mortality studies. Although they have limitations related to study size (Chiou et al. 2001) or control-selection methods (Ferreccio et al. 2000), these studies represent valuable contributions to the epidemiological database that addresses cancer risk from arsenic in drinking water. As with the data from southwestern Taiwan, the difference between the exposure concentrations in these studies and the concentrations of current regulatory concern is relatively small. Therefore, the range of concentrations for which the dose-response curve must be extrapolated is very small.

• The findings of mortality in the arsenic endemic area of southwestern Taiwan in the large ecological study of Tsai et al. (1999) indicate that use of the regional and national rates as referents for mortality studies in this region is appropriate and that important confounding is unlikely when these external rates are incorporated in a quantitative risk assessment.

• Kurttio et al. (1999) demonstrated increased bladder cancer risks in smokers at very low concentrations of arsenic in drinking water. Because the study was small, the very high risks generated by Kurttio et al. (1999) are not as concordant as the risks that emerge from the studies in southwestern Taiwan, northeastern Taiwan, Chile, and Argentina. The possibility exists that the study was distinguished by some unmeasured bias or that the findings occurred by chance.

• Several limitations of the study by Lewis et al. (1999a), including issues surrounding the exposure metric and differences between the study population and the control population, preclude its use for hazard evaluation and risk assessment.

• The studies by Karagas et al. (2001) and Tucker et al. (2001) confirmed the association between exposure to arsenic in drinking water and skin cancer. The former study is complicated by the use of an exposure measure (arsenic in toenails) with poorly understood toxicokinetics and significant opportunity for misclassification due to external contamination. The latter study reported few cases and used exposure metrics that were difficult to interpret, limiting its utility for modeling dose-response relationship. Finally, the subcommittee notes that internal cancers are more appropriate as an end point for risk assessment than nonmelanomic skin cancer, because internal cancers are more likely to be life-threatening.

• Overall, the data from southwestern Taiwan (Chen et al 1985, 1992; Wu et al. 1989) remain as the preferred data for use in quantitative risk assessment. The data from Chiou et al. (2001) and Ferreccio et al. (2000) can be modeled to augment analyses of the southwestern Taiwanese data. From a public-health perspective, the database of epidemiological studies linking arsenic in drinking water with increased risk of skin, bladder, and lung cancer provides a sound and adequate basis for quantitative assessment of cancer risk.

RECOMMENDATIONS

• Epidemiological studies are highly recommended to investigate the dose-response relationship between arsenic ingestion and the noncancer end points of circulatory system disease (particularly hypertension, cardiovascular disease, and cerebrovascular disease) and diabetes mellitus. Because these end points are common causes of morbidity and mortality, even small potential increases in relative risk at low arsenic doses could be of considerable public-health importance. Laboratory and clinical studies should investigate the modes of action of arsenic for these end points.

• Epidemiological studies are recommended to further investigate the relationship between arsenic ingestion and adverse reproductive outcomes.

• Studies are needed to define the pathological features of a potential link between arsenic ingestion and respiratory function. A possible impact of arsenic exposure on neuro-cognitive development in children requires further investigation. As noted in the 1999 NRC report, the effect of arsenic on immune function also merits further study.

• There is a need for additional epidemiological research in other populations from other geographic areas. In future epidemiological studies that investigate both cancer and noncancer outcomes of ingestion of arsenic in drinking water, detailed information on exposure should be collected. This information should include water-ingestion rates and intake of food containing arsenic, as well as long-term histories of water sources and arsenic concentrations associated with those sources. As was noted in the 1999 NRC report, data related to latency and the relationship between magnitude of dose and time course of exposure should be obtained.

• Future studies should consider interactions with host factors that might influence susceptibility to the adverse effects of arsenic exposure, including age at exposure, dietary status, smoking, and genetic polymorphisms that could affect arsenic metabolism.

• Epidemiological studies should be designed to be of sufficient size to determine risk in subpopulations that might be susceptible to the adverse effects of arsenic and to quantify the extent of interaction with host factors.

• Data from additional follow-up of exposed populations, including populations in Utah, Chile, Bangladesh and West Bengal, should be considered, as appropriate, in future risk assessments. Exploring different dose metrics in further analyses of the data from Utah is also warranted.

REFERENCES

Ahmad, S., W.L. Anderson, and K.T. Kitchin. 1999. Dimethylarsinic acid effects on DNA damage and oxidative stress related biochemical parameters in B6C3F1 mice. Cancer Lett. 139(2):129-135.

Ahmad, S., M.H. Sayed, S. Barua, M.H. Khan, M.H. Faruquee, A. Jalil, S.A. Hadi, and H.K. Talukder. 2001. Arsenic in drinking water and pregnancy outcomes. Environ. Health Perspect. 109(6):629-631.

Ahsan, H., M. Perrin, A. Rahman, F. Parvez, M. Stute, Y. Zheng, A.H. Milton, P. Brandt-Rauf, A. van Geen, and J. Graziano. 2000. Associations between drinking water and urinary arsenic levels and skin lesions in Bangladesh. J. Occup. Environ. Med. 42(12):1195-1201.

Armenian, H.K.. 1987. Incubation periods of cancer: old and new. J. Chronic Dis. 40 (suppl. 2):9S-15S.

Bates, M.N., A.H. Smith, and K.P. Cantor. 1995. Case-control study of bladder cancer and arsenic in drinking water. Am. J. Epidemiol. 141(6):523-530.

Boonchai, W., A. Green, J. Ng, A. Dicker, and G. Chenevix-Trench. 2000. Basal cell carcinoma in chronic arsenicism occurring in Queensland, Australia, after ingestion of an asthma medication. J. Am. Acad. Dermatol. 43(4):664-669.

Buchet, J.P., and D. Lison. 1998. Mortality by cancer in groups of the Belgian population with a moderately increased intake of arsenic. Int. Arch. Occup. Environ. Health 71(2):125-130.

Calderon, J., M.E. Navarro, M.E. Jimenez-Capdeville, M.A. Santos-Diaz, A. Golden, I. Rodriguez-Leyva, V. Borja-Aburto, and F. Diaz-Barriga. 2001. Exposure to arsenic and lead and neuropsychological development in Mexican children. Environ. Res. 85(2):69-76.

Cantor, K.P. 2001. Invited commentary: Arsenic and cancer of the urinary tract. Am. J. Epidemiol. 153(5):422-423.

CDC (Center for Disease Control). 2001. International Classification of Diseases, Ninth Revision (ICD-9). Mortality Data from the National Vital Statistics System. National Center for Health Statistics. [Online]. Available: www.cdc.gov/nchs/ about/ major/ dvs/ icd9des.html. [August 4, 2001].

Chen, C.J., and H.Y. Chiou. 2001. Chen and Chiou respond to "Arsenic and cancer of the urinary tract" by Cantor. Am. J. Epidemiol. 153(5):422-423.

Chen, C.J., and C.J. Wang. 1990. Ecological correlation between arsenic level in well water and age-adjusted mortality from malignant neoplasms. Cancer Res. 50(17):5470-5474.

Chen, C.J., C.W. Chen, M.M. Wu, and T.L. Kuo. 1992. Cancer potential in liver, lung, bladder, and kidney due to ingested inorganic arsenic in drinking water. Br. J. Cancer 66(5):888-892.

Chen, C.J., Y.C. Chuang, T.M. Lin, and H.Y. Wu. 1985. Malignant neoplasms among residents of a blackfoot disease-endemic area in Taiwan: high-arsenic artesian well water and cancers. Cancer Res. 45(11 Pt 2):5895-5899.

Chen, C.J., Y.C. Chuang, S.L. You, T.M. Lin, and H.Y. Wu. 1986. A retrospective study on malignant neoplasms of bladder, lung and liver in blackfoot disease endemic area of Taiwan. Br. J. Cancer 53(3):399-405.

Chen, C.J., Y.M. Hsueh, M.S. Lai, M.P. Shyu, S.Y. Chen, M.M. Wu, T.L. Kuo, and T.Y. Tai. 1995. Increased prevalence of hypertension and long-term arsenic exposure. Hypertension 25(1):53-60.

Chiou, H.Y., S.T. Chiou, Y.H. Hsu, Y.L. Chou, C.H. Tseng, M.L. Wei, and C.J. Chen. 2001. Incidence of transitional cell carcinoma and arsenic in drinking water: a follow-up study of 8,102 residents in an arseniasis-endemic area in northeastern Taiwan. Am. J. Epidemiol. 153(5):411-418.

Chiou, H.Y., Y.M. Hsueh, K.F. Liaw, S.F. Horng, M.H. Chiang, Y.S. Pu, J.S. Lin, C.H. Huang, and C.J. Chen. 1995. Incidence of internal cancers and ingested inorganic arsenic: a seven-year follow-up study in Taiwan. Cancer Res. 5:1296-1300.

Chowdhury, U.K., B.K. Biswas, T.R. Chowdhury, G. Samanta, B.K. Mandal, G.C. Basu, C.R. Chanda, D. Lodh, K.C. Saha, S.K. Mukherjee, S. Roy, S. Kabir, Q. Quamruzzaman, and D. Chakraborti. 2000. Groundwater arsenic contamination in Bangladesh and West Bengal, India. Environ. Health Perspect. 108(5):393-397.

Concha, G.G. Vogler, D. Lezcano, B. Nernell, and M. Vahter. 1998. Exposure to inorganic arsenic metabolites during early human development. Toxicol. Sci. 44(2):185-190.

Cuzick, J., S. Evans, M. Gillman, and D.A. Price Evans. 1982. Medicinal arsenic and internal malignancies. Br. J. Cancer 45(6):904-911.

Cuzick, J., P. Sasieni, and S. Evans. 1992. Ingested arsenic, keratoses, and bladder cancer. Am. J. Epidemiol. 136(4):417-421.

EPA (U.S. Environmental Protection Agency). 2000. A Re-Analysis of Arsenic-Related Bladder and Lung Cancer Mortality in Millard County, Utah. EPA 815-R-00-027. Office of Water, U.S. Environmental Protection Agency, Washington, D.C.

Ferreccio, C., C. Gonzalez, V. Milosavljevic, G. Marshall, A.M. Sancha, and A.H. Smith. 2000. Lung cancer and arsenic concentrations in drinking water in Chile. Epidemiology 11(6):673-679.

Fierz, U. 1965. Follow-up studies of the side-effects of the treatment of skin diseases with inorganic arsenic. [in German]. Dermatologica 131(1):41-58.

Garcia-Vargas, G.G., L.M. Del Razo, M.E. Cebrian, A. Albores, P. Ostrosky-Wegman, R. Montero, M.E. Gonsbatt, C.K. Lim, and F. De Matteis. 1994. Altered urinary porphyrin excretion in a human population chronically exposed to arsenic in Mexico. Hum. Exp. Toxicol. 13(12):839-847.

Guo, H.R., H.S. Chiang, H. Hu, S.R. Lipsitz, and R.R. Monson. 1997. Arsenic in drinking water and incidence of urinary cancers. Epidemiology 8(5):545-550.

Hernandez-Zavala, A., L.M. Del Razo, C. Aguilar, G.G. Garcia-Vargas, V.H. Borja, and M.E. Cebrian. 1998. Alteration in bilirubin excretion in individuals chronically exposed to arsenic in Mexico. Toxicol. Lett. 99(2):79-84.

Hernandez-Zavala, A., L.M. Del Razo, G.G. Garcia-Vargas, C. Aguilar, V.H. Borja, A. Albores, and M.E. Cebrian. 1999. Altered activity of heme biosynthesis pathway enzymes in individuals chronically exposed to arsenic in Mexico. Arch. Toxicol. 73(2):90-95.

Hertz-Picciotto, I., H.M. Arrighi, and S.W. Hu. 2000. Does arsenic exposure increase the risk for circulatory disease? Am. J. Epidemiol. 151(2):174-181.

Hinwood, A.L., D.J. Jolley, and M.R. Sim. 1999. Cancer incidence and high environmental arsenic concentrations in rural populations: results of an ecological study. Int. J. Environ. Health Res. 9(2):131-141.

Hopenhayn-Rich, C., M.L. Biggs, A. Fuchs, R. Bergoglio, E.E. Tello, H. Nicolli, and A.H. Smith. 1996. Bladder cancer mortality associated with arsenic in drinking water in Argentina. Epidemiology 7(2):117-124.

Hopenhayn-Rich, C., M.L. Biggs, and A.H. Smith. 1998. Lung and kidney cancer mortality associated with arsenic in drinking water in Córdoba, Argentina. Int. J. Epidemiol. 27(4):561-569.

Hopenhayn-Rich, C., S.R. Browning, I. Hertz-Picciotto, C. Ferreccio, C. Peralta, and H. Gibb. 2000. Chronic arsenic exposure and risk of infant mortality in two areas of Chile. Environ. Health Perspect. 108(7):667-673.

Ihrig, M.M., S.L. Shalat, and C. Baynes. 1998. A hospital-based case-control study of stillbirths and environmental exposure to arsenic using an atmospheric dispersion model linked to a geographical information system. Epidemiology 9(3):290-294.

IOM (Institute of Medicine). 2001. Dietary Reference Intakes for Vitamin A, Vitamin K, Arsenic, Boron, Chromium, Copper, Iodine, Iron, Manganese, Molybdenum, Nickel, Silicon, Vanadium, and Zinc. Washington, DC: National Academy Press.

Karagas, M.R., T.D. Tosteson, J. Blum, B. Klaue, J.E. Weiss, V. Stannard, V. Spate, and J.S. Morris. 2000. Measurement of low levels of arsenic exposure: a comparison of water and toenail concentrations. Am. J. Epidemiol. 152(1):84-90.

Karagas, M.R., T.A. Stukel, J.S. Morris, T.D. Tosteson, J.E. Weiss, S.K. Spencer, and E.R. Greenberg. 2001. Skin cancer risk in relation to toenail arsenic concentrations in a US population-based case-control study. Am. J. Epidemiol. 153(6):559-565.

Kurokawa, M., K. Ogata, M. Idemori, S. Tsumori, H. Miyaguni, S. Inoue, N. Hotta, and Nobuyuki. 2001. Investigation of skin manifestations of arsenicism due to intake of arsenic-contaminated groundwater in residents of Samta, Jessore, Bangladesh. Arch. Dermatol. 137(1):102-103.

Kurttio, P., E. Pukkala, H. Kahelin, A. Auvinen, and J. Pekkanen. 1999. Arsenic concentrations in well water and risk of bladder and kidney cancer in Finland. Environ. Health Perspect. 107(9):705-710.

Lai, M.S., Y.M. Hsueh, C.J. Chen, M.P. Shyu, S.Y. Chen, T.L. Kuo, M.M. Wu, and T.Y. Tai. 1994. Ingested inorganic arsenic and prevalence of diabetes mellitus. Am. J. Epidemiol. 139(5):484-492.

Lewis, D.R., J.W. Southwick, R. Ouellet-Hellstrom, J. Rench, and R.L. Calderon. 1999a. Drinking water arsenic in Utah: a cohort mortality study. Environ. Health Perspect. 107(5):359-365.

Lewis, D.R., R.L. Calderon, J.W. Southwick, R. Ouellet-Hellstrom, and J. Rench. 1999b. "Drinking water arsenic in Utah ...". Response. Environ. Health Perspect. 107(11):A544-546.

Ma, H.Z., Y.J. Xia, K.G. Wu, T.Z. Sun, and J.L. Mumford. 1999. Human exposure to arsenic and health effects in Bayingnormen, Inner Mongolia. Pp. 127-131 in Arsenic Exposure and Health Effects, W.R. Chappell, C.O. Abernathy, and R.L. Calderon, eds. Oxford: Elsevier.

Mazumder, D.N., J.D. Gupta, A. Santra, A. Pal, A. Ghose, and S. Sarkar. 1998. Chronic arsenic toxicity in West Bengal—The worst calamity in the world. J. Indian Med. Assoc. 96(1):4-7,18.

Mazumder, D.N., R. Haque, N. Ghosh, B.K. De, A. Santra, D. Chakraborti, and A.H. Smith. 2000. Arsenic in drinking water and the prevalence of respiratory effects in West Bengal, India. Int. J. Epidemiol. 29(6):1047-1052.

Milton, A.H., Z. Hasan, A. Rahman, and M. Rahman. 2001. Chronic arsenic poisoning and respiratory effects in Bangladesh. J. Occup. Health 43(3):136-140.

Morales, K.H., L. Ryan, T.L. Kuo, M.M. Wu, and C.J. Chen. 2000. Risk of internal cancers from arsenic in drinking water. Environ. Health Perspect. 108(7):655-661.

Neubauer, O. 1947. Arsenical cancer: A review. Br. J. Cancer 1(June):192-251.

NRC (National Research Council). 1999. Arsenic in Drinking Water. Washington, DC: National Academy Press.

Rahman, M., M. Tondel, S.A. Ahmad, and O. Axelson. 1998. Diabetes mellitus associated with arsenic exposure in Bangladesh. Am. J. Epidemiol. 148(2):198-203.

Rahman, M., M. Tondel, S.A. Ahmad, I.A. Chowdhury, M.H. Faruquee, and O. Axelson. 1999. Hypertension and arsenic exposure in Bangladesh. Hypertension 33(1):74-78.

Santra, A., J. Das Gupta, B.K. De, B. Roy, and D.N. Mazumder. 1999. Hepatic manifestations in chronic arsenic toxicity. Indian J. Gastroenterol. 18(4):152-155.

Siripitayakunkit, P., M. Visudhiphan, M. Pradipasen, and T. Vorapongsathron. 1999. Association between chronic arsenic exposure and children's intelligence in Thailand. Pp. 141-150 in Arsenic Exposure and Health Effects, W.R. Chappell, C.O. Abernathy, and R.L. Calderon, eds. New York: Elsevier.

Smith, A.H., M. Goycolea, R. Haque, and M.L. Biggs. 1998. Marked increase in bladder and lung cancer mortality in a region of northern Chile due to arsenic in drinking water. Am. J. Epidemiol. 147(7):660-669.

Sommers, S.G., and R.G. McManus. 1953. Multiple arsenical cancers of skin and internal organs. Cancer 6:347-359.

Southwick, J.W., A.E. Western, M.M. Beck, J. Whitley, and R. Isaacs. 1982. Community Health Associated with Arsenic in Drinking Water in Millard County, Utah. EPA-600/S1-81-064. Health Effects Research Laboratory, U.S. Environmental Protection Agency, Cincinnati, OH.

Tondel, M., M. Rahman, A. Magnuson, I.A. Chowdhury, M.H. Faruquee, and S.A. Ahmad. 1999. The relationship of arsenic levels in drinking water and the prevalence rate of skin lesions in Bangladesh. Environ. Health Perspect. 107(9):727-729.

Tsai, S.-M., T.-N. Wang, and Y.-C. Ko. 1998. Cancer mortality trends in a blackfoot disease endemic community of Taiwan following water source replacement. J. Toxicol. Environ. Health Part A 55(6):389-404.

Tsai, S.-M., T.-N. Wang, and Y.-C. Ko. 1999. Mortality for certain diseases in areas with high levels of arsenic in drinking water. Arch. Environ. Health 54(3):186-193.

Tseng, C.H., T.Y. Tai, C.K. Chong, C.P. Tseng, M.S. Lai, B.J. Lin, H.Y. Chiou, Y.M. Hsueh, K.H. Hsu, and C.J. Chen. 2000. Long-term arsenic exposure and incidence of non-insulin-dependent diabetes mellitus: a cohort study in arseniasis-hyperendemic villages in Taiwan. Environ. Health Perspect. 108(9):847-851.

Tsuda, T., A. Babazono, E. Yamamoto, N. Kurumatani, Y. Mino, T. Ogawa, Y. Kishi, and H. Aoyama. 1995. Ingested arsenic and internal cancer: a historical cohort study followed for 33 years. Am. J. Epidemiol. 141(3):198-209.

Tucker, S.B., S.H. Lamm, F.X. Li, R. Wilson, D.M. Byrd, S. Lai, Y. Tong, and L. Loo. 2001. Relationship Between Consumption of Arsenic-Contaminated Well Water and Skin Disorders in Huhhot, Inner Mongolia. Inner Mongolia Cooperative Arsenic Project (IMCAP) Study. Final Report. Department of Dermatology,

University of Texas, Houston, and Department of Environmental Epidemiology, Institute of Environmental Engineering, Chinese Academy of Preventive Medicine, Beijing, China. July 5, 2001.

Wang, S.L., W.H. Pan, C.M. Hwu, L.T. Ho, C.H. Lo, S.L. Lin, and Y.S. Jong. 1997. Incidence of NIDDM and the effects of gender, obesity, and hyperinsulinaemia in Taiwan. Diabetologia 40(12):1431-1438.

Wong, S.S., K.C. Tan, and C.L. Goh. 1998. Cutaneous manifestations of chronic arsenicism: review of seventeen cases. J. Am. Acad. Dermatol. 38(2 Pt 1):179-185.

Wu, M.M., T.L. Kuo, Y.H. Hwang, and C.J. Chen. 1989. Dose-response relation between arsenic concentration in well water and mortality from cancers and vascular diseases. Am. J. Epidemiol. 130(6):1123-1132.

3

Experimental Studies

This chapter discusses the effects of arsenic that have been observed in experimental studies. It begins with a summary of the experimental studies described in the 1999 report. Following that summary, toxicokinetic, animal toxicity, and mechanistic studies of arsenic, published since the previous NRC report was released, are discussed. This chapter does not provide a comprehensive discussion of all the toxicologic mechanisms of arsenic.

SUMMARY OF EXPERIMENTAL STUDIES
DISCUSSED IN THE 1999 REPORT

The previous NRC Subcommittee on Arsenic in Drinking Water reviewed the data on the toxicokinetics, animal toxicity studies, and mode of action of arsenic, focusing on the modes of action that might underlie the carcinogenic effects of arsenic (NRC 1999). It concluded that inorganic arsenic is readily absorbed from the gastrointestinal tract in humans and it is mainly transported in the blood bound to sulfhydryl groups. At low-to-moderate doses, inorganic arsenic has a half-life in the body of about 4 days and is excreted primarily in the urine (NRC 1999). Humans and some animals methylate inorganic arsenic compounds to pentavalent monomethylarsonic acid (MMA^V) and pentavalent dimethylarsinic acid (DMA^V), which are less acutely toxic and readily excreted. At the time of the NRC 1999 report, there was little information on the distribution and toxicity of the trivalent methylated metabolites (monomethyl-arsonous acid (MMA^{III}) and dimethylarsinous acid (DMA^{III})). It was noted that the fractions of the various metabolites of arsenic in urine (inorganic arsenic, MMA, and DMA) vary markedly among humans, and the toxicokinetics of arsenic varies considerably among animal species. It is not known in which animal species the toxicokinetics more closely resembles that in humans, and this uncertainty makes it difficult to extrapolate from animals to humans.

The subcommittee concluded that the mechanisms or modes of action by which inorganic arsenic causes toxicity, including cancer, is not well established (NRC 1999). The data available on the ability of inorganic arsenic to act as a cocarcinogen or a tumor promoter in rats and mice are conflicting, but studies conducted at very high doses indicate that DMA^{III} is not a tumor initiator but might act as a tumor promoter. Furthermore, inorganic arsenic and its metabolites have been shown to induce chromosomal alterations (aberrations, aneuploidy, and sister chromatid exchange) and large deletion mutations, but not point mutations. Data on other genotoxic responses that might indicate mode of action for arsenic were not sufficient for conclusions to be drawn. Therefore, the 1999 subcommittee concluded that "the most plausible and generalized mode of action for arsenic carcinogenicity is that it induces structural and numerical chromosomal abnormalities without acting directly with DNA." The subcommittee also discussed other mechanisms, such as cell proliferation and oxidative stress. An indirect mechanism of mutagenicity suggests that the most plausible shape of the carcinogenic dose-response curve is sublinear "at some point below the level at which a significant increase in tumors is observed [in the available epidemiological studies]." There was insufficient scientific evidence to identify the dose at which sublinearity might occur. Therefore, the subcommittee concluded that "because a specific mode (or modes) of action has not been identified at this time, it is prudent not to rule out the possibility of a linear response."

The subcommittee further concluded that arsenicals inhibit some types of mitochondrial-respiratory function, leading to decreased cellular ATP production and increased production of hydrogen peroxide (H_2O_2) (NRC 1999). Those effects could cause the formation of reactive oxygen species, resulting in oxidative stress. Oxidative stress can have numerous effects, including inhibition of heme-biosynthetic pathways and induction of major stress proteins. Although the role of arsenic-induced oxidative stress in mediating DNA damage is not clear, the intracellular production of reactive oxygen species might play an initiating role in the carcinogenic process by producing DNA damage. In the remainder of this chapter, more recent studies on the toxicokinetics of arsenic will be discussed, followed by in vitro and in vivo studies that provide additional information on the mode of action of arsenic.

TOXICOKINETICS

Methylation of Arsenic

Arsenic can exist in methylated and inorganic forms as well as in different valence states (NRC 1999). The form and valence state can affect the toxicity of arsenic; therefore, it is important to understand the metabolism and toxicokinetics of arsenic.

As discussed in the 1999 NRC report, inorganic arsenic is believed to be methylated via sequential reduction of pentavalent arsenic to trivalent arsenic, followed by oxidative addition of a methyl group from S-adenosylmethionine (SAM) to the trivalent form (Figure 3-1). The main products of that methylation, MMA^V and DMA^V, are readily excreted in the urine. More recent experiments have detected the presence of the reduced methylated forms (MMA^{III} and DMA^{III}) in human urine.[1] The development of the analytical methods for the speciation of arsenic metabolites, as well as the advantages and disadvantages of these methods, were thoroughly discussed in the previous NRC report (NRC 1999). The reduced methylated forms and their toxicity are discussed later in this chapter.

Further methylation of DMA to trimethylarsine is frequently seen in microorganisms exposed to arsenite (NRC 1999). A small percentage of urinary arsenic as trimethylarsine oxide (TMAO) has been detected in mice, hamsters, and humans following exposure to DMA (for review, see Kenyon and Hughes 2001), but TMAO or demethylated products of DMA were not detected in the blood or tissues of mice exposed intravenously to DMA at a dose of 1 or 100 mg/kg (Hughes et al. 2000). By contrast, TMAO has not been reported to be present in the urine of mammals exposed to inorganic arsenic. DMA formed by methylation of inorganic arsenic (As^{III}) and MMA^{III} has been shown to clear

---

[1]Because trivalent methylated forms of arsenic have been detected only recently, they often have not been specifically assayed for or discussed in most experiments. In those cases, the abbreviations MMA and DMA are used without indicating the valence state.

rapidly from cells (Styblo et al. 1999a; Lin et al. 2001). That rapid clearance might prevent the accumulation of intracellular concentrations of DMA required for further methylation to TMAO, explaining why TMAO is not detected following exposure to As^{III}.

Nonenzymatic methylation of arsenic has been seen in vitro. Studies have demonstrated that methylcobalamin (a form of vitamin B_{12}) can mediate the nonenzymatic methylation of arsenite (Zakharyan and Aposhian 1999a), but whether that occurs in vivo is not known.

Glutathione (GSH), and possibly other thiols, can act as reducing agents in the methylation process (NRC 1999). Recent in vitro studies indicate that dithiols (e.g., reduced lipoic acid) might be more active than GSH in providing the reducing environment required for methylation by MMA^{III} methyltransferase (Zakharyan et al. 1999). Thiol binding might also be involved in arsenic biotransformation. Trivalent arsenic metabolites are highly bound to cytosolic proteins in the cells (Styblo and Thomas 1997). In in vitro studies, protein-bound inorganic arsenic and MMA were methylated to MMA and DMA, respectively. It has been proposed that arsenic is bound to a protein dithiol cofactor before the sequential methylation of arsenic (Thompson 1993; DeKimpe et al. 1999a).

The exact sequence of events in arsenic biotransformation remains unknown, but it seems clear that S-adenosylmethionine (SAM) is the main source of methyl

$$H_2AsO_4^- \xrightarrow[\text{Reductase}]{2e^-} AsO_3^{3-} \xrightarrow[\substack{\text{Methyl-}\\\text{transferase}}]{CH_3^+} \underset{MMA^V}{CH_3AsO_3^{2-}} \xrightarrow[\text{Reductase}]{2e^-} \underset{MMA^{III}}{CH_3AsO_2^{2-}} \rightarrow$$

$$\xrightarrow[\substack{\text{Methyl-}\\\text{transferase}}]{CH_3^+} \underset{DMA^V}{(CH_3)_2AsO_2^-} \xrightarrow[\text{Reductase}]{2e^-} \underset{DMA^{III}}{(CH_3)_2AsO^-}$$

FIGURE 3-1 Proposed chemical pathway for the methylation of inorganic arsenic in humans. The enzymes that have been proposed for the reduction and methylation reactions are indicated. Uncertainties regarding this pathway are discussed in the text.

groups for the methylation of arsenic and that the methylation is dependent upon methyltransferases. Studies have shown that SAM is required for arsenic methylation in various in vitro systems and that arsenic methylation in vivo is decreased by specific inhibitors of SAM-dependent methylation and by low methionine intake (NRC 1999). Because of the importance of SAM in the methylation of arsenic, much research has been aimed at characterizing the methyltransferases involved in SAM-dependent arsenic methylation. During recent years, arsenite and MMA[III] methyltransferases from the liver of rabbits, hamsters, and rhesus monkeys have been purified and partially characterized (Wildfang et al. 1998; Zakharyan et al. 1999). The K_m, determined using Michaelis-Menten kinetics, for arsenite methyltransferase for hamsters was 1.79×10^{-6} M and for MMA methyltransferase was 7.98×10^{-4} M. The MMA methyltransferase was higher than the arsenite methyltransferase. Similar values were reported for rabbit methyltransferases, but the rhesus monkey had K_m values for MMA methyltransferase of 3.5×10^{-6} M and for arsenite methyltransferase of 5.5×10^{-6} M. Zakharyan et al. (1999) reported that the rabbit liver MMA methyltransferase had higher affinity for MMA[III] than for MMA[V]. Furthermore, the K_m for MMA[III] methyltransferase from Chang human hepatocytes was not very different from that of rabbit liver.

The exact structures of the arsenic methyltransferases have not been determined (NRC 1999), but recent in vitro studies using rabbit liver cytosol showed that the two steps in arsenic methylation to DMA are markedly inhibited by pyrogallol (0.3 to 9 millimolar) (mM), a specific inhibitor of catechol-*O*-methyltransferase (DeKimpe et al. 1999a). Those data suggest that the active sites of arsenite methyltransferase and MMA methyltransferase are similar to that of catechol-*O*-methyltransferase. Furthermore, trichloromethiazide (TCM), an inhibitor of human microsomal thiopurine methyltransferase, did not inhibit the formation of MMA but did inhibit the formation of DMA. In contrast, neither of those arsenic methylation steps was inhibited by an inhibitor of cytosolic thiopurine methyltransferase (*p*-anisic acid) or by an inhibitor of cytosine DMA methyltransferase.

A DMA methyltransferase, which would further methylate DMA forming TMAO, has not been reported to be present in mammalian cells. Styblo et al. (1999b) reported that DMA was the only metabolite detected in rat or human hepatocytes incubated with DMA[V] or a glutathione complex of DMA[III] (DMA[III]–GS).

Species Differences in the Methylation of Arsenic

As discussed in the previous report, there is considerable variation in the methylation of inorganic arsenic among mammalian species (NRC 1999). Rats, mice, and dogs show a very efficient methylation of arsenic to DMA. Rabbits and hamsters also methylate arsenic relatively efficiently. In most animals, the DMA that is formed is rapidly excreted in urine. However, in rats, most of the DMA that is formed accumulates in the red blood cells and tissues. Rats also appear to methylate administered DMA to TMAO more efficiently than other species (Kitchin et al. 1999; NRC 1999). Following exposure of male rats to high concentrations of DMA in drinking water (100 mg/L, equivalent to 100,000 µg/L), about 10% of the urinary arsenic was present as TMAO (Yoshida et al. 1998).

Consistent with that marked variation in arsenic methylation efficiency seen among animal species (Vahter 1999b), the activity of methyltransferases differs markedly among the animals studied (Healy et al. 1999). The variation in the activity of those enzymes probably underlies most of the cross-species variability in methylation ability (for review, see Healy et al. 1999; Vahter 1999a). Guinea pigs and several types of non-human primates, including the chimpanzee, seem to be unable to methylate inorganic arsenic (Healy et al. 1999; Vahter 1999b), and no methyltransferase activity was detected in those species (Healy et al. 1999).

Human arsenic methyltransferases were long thought to be very unstable, because activity could not be detected in human liver preparations (NRC 1999). Recently, however, arsenic-methylation activity was detected in human hepatocytes; MMA^{III} methyltransferase activity was detected in cultured Chang human hepatocytes (Zakharyan et al. 1999). Incubation of primary human hepatocytes with arsenite yielded MMA and DMA, and incubation with MMA^{III} produced mainly DMA (Styblo et al. 1999a). Rat hepatocytes, however, methylated arsenic considerably faster than did human hepatocytes (Styblo et al. 1999a; Styblo et al. 2000).

In contrast to humans, most mammals do not excrete appreciable amounts of MMA in the urine. Recently, however, the Flemish Giant rabbit was found to excrete substantial amounts of MMA in urine (DeKimpe et al. 1999b), and MMA was formed in vitro after incubation of inorganic arsenic with rabbit-liver cytosol (DeKimpe et al. 1999a).

No differences, however, were seen in the urinary pattern of arsenic metabolites in three strains of mice 24 hours (hr) after administration of an oral dose of arsenite (Hughes et al. 1999). More than 95% of the arsenic in urine (corresponding to 60%, 68%, and 69% of the administered dose in each of the three mouse strains) was in the form of DMA in all three mouse strains.

As indicated above, the ability to excrete MMA and DMA in the urine varies among species. Because of this variation and other species differences in arsenic metabolism discussed in this section, extrapolation of data from studies in animals and animal cells to humans is difficult.

Tissue Differences in the Methylation of Arsenic

The liver appears to play the central role in arsenic methylation (NRC 1999), but in vitro studies using male mouse cytosol demonstrated that most tissues appear to be capable of methylating arsenic (Healy et al. 1998). The highest activity of arsenite methyltransferase was observed in the cytosol from the testis, followed by cytosol from the kidney, liver, and lung. More recent data also point to the liver's important role. Styblo et al. (2000) reported a much higher rate of arsenic methylation in primary human hepatocytes compared with human keratinocytes and bronchial cells, and no methylation activity was detected in human urinary-bladder cells. When proliferating human keratinocytes and bronchial epithelial cells were cultured in the presence of the relatively low arsenite concentration of 0.05 micromolar (μM) (approximately 3.7 μg/L), more than two-thirds of the cell-associated methylated arsenic consisted of MMA, most of which was retained intracellularly throughout a 24-hr incubation. Human keratinocytes cultured in the presence of 1 μM of methylarsine oxide, a putative substrate for MMA^{III} methyltransferase, did not produce any DMA. Recent studies indicate that MMA^{III} might be the most toxic intracellular form of arsenic in terms of oxidative stress, enzyme inhibition, and DNA damage (see Mechanistic Data later in this chapter). It is noteworthy that cells from two tissues that are targets of arsenic-induced cancer (skin and lung) seem to have less efficient conversion of MMA to DMA at relevant concentrations of arsenite in culture—that is, at concentrations similar to those that might occur in blood and possibly tissue following chronic ingestion of low-to-moderate concentrations of arsenic. However, more studies on this topic are needed before firm conclusions can be reached.

The situation is even more complex in vivo, where the extent of methylation of inorganic arsenic to MMA and DMA is also influenced by the rate of cellular uptake of the inorganic arsenic in the various tissues. Tatum and Hood (1999) reported that the uptake and methylation of As^{III} by a kidney-epithelium-derived cell line, the NKR-52E cell line, were lower than those of primary rat hepatocytes or hepatoma-derived cell lines. Those data confirm previous findings that As^{III} is the main form of arsenic taken up by the liver (NRC 1999). In the presence of phosphate-free media, the uptake and cytotoxicity of As^V in KB oral epidermoid carcinoma cells was greatly enhanced (Huang and Lee 1996).

Cellular uptake and efflux of arsenic vary considerably among the different arsenic metabolites, and the variation affects the distribution of metabolites formed in the liver following absorption of arsenic compounds. In comparison to the trivalent methylated forms of arsenic, exposure to the pentavalent forms (MMA^V or DMA^V) results in very low tissue concentrations of MMA and DMA (Hughes and Kenyon 1998; NRC 1999). That difference is probably because of a lower cellular uptake and accumulation of the pentavalent forms than the trivalent forms. A lower uptake was confirmed in studies in rat and human hepatocytes that showed a several-fold higher cellular uptake of As^{III} and MMA^{III} than of the corresponding pentavalent forms (Styblo et al. 1999a). In the presence of phosphate-free media, the uptake and cytotoxicity of As^V in KB oral carcinoma cells were greatly enhanced (Huang and Lee 1996).

Cellular efflux also appears to vary among the different forms of arsenic. Styblo et al. (1999a) demonstrated that DMA is the main excretory product in human and rat hepatocytes. Inhibition of the methylation of MMA to DMA by increasing arsenite concentrations in the medium resulted in the accumulation of MMA in the cells, indicating that MMA is not excreted as readily from liver cells as is DMA. Whether that same effect occurs in cells from other tissues is unknown.

Induction of Arsenic Methyltransferases

The inducibility of arsenic methyltransferase has been investigated in mice. Arsenic methyltransferase activity did not appear to be induced in mice exposed subchronicly to arsenic in the drinking water (25 or 2,500 μg/L) (Hughes and Thompson 1996; Healy et al. 1998).

Trivalent Methylated Arsenic Metabolites

It is obvious from the proposed scheme of arsenic methylation (Figure 3-1) that pentavalent methylated arsenic compounds can be reduced to trivalent methylated arsenic compounds (MMA^{III} and DMA^{III}). Although pentavalent arsenicals can be reduced directly, for example by glutathione (NRC 1999), recent studies indicate the involvement of arsenic reductases. Arsenate reductase activity has been detected in human liver (Radabaugh and Aposhian 2000). The enzyme has a molecular weight of about 72 kilodaltons (kDa), and it requires a thiol and a heat-stable cofactor for activity. It did not reduce MMA^V, indicating the presence of two enzymes, arsenate reductase and MMA^V reductase.

MMA^V reductase activity has been detected in rabbit liver (Zakharyan and Aposhian 1999b), hamster tissues (Sampayo-Reyes et al. 2000), and human liver (Zakharyan et al. 2001). There is evidence that the human MMA^V reductase is identical to glutathione-*s*-transferase omega class 1-1. The rabbit liver MMA^V reductase was shown to reduce both DMA^V and arsenate (As^V) (Zakharyan and Aposhian 1999b). The K_m values were 2.16×10^{-3} M with MMA^V as the substrate, 20.9×10^{-3} M with DMA^V as the substrate, and 109×10^{-3} M with arsenate as the substrate. When the K_m for the rabbit liver MMA^V reductase was compared with that of the As^{III} methyltransferase (i.e., 5.5×10^{-6}) and that of MMA^{III} methyltransferase (i.e., 9.2×10^{-6}), the authors concluded that MMA^V reductase was the rate-limiting enzyme for arsenite metabolism in the rabbit liver. However, it should be emphasized that, because of the species differences in arsenic methylation (see above), it is not known whether the rate-limiting step in the rabbit liver equates to the rate-limiting step in any particular human organ.

It is not clear to what extent the DMA formed following exposure to inorganic arsenic is reduced to DMA^{III} by a specific DMA^V reductase. The latter has not been well studied, but DMA^{III} was detected in the liver of hamsters given arsenate (Sampayo-Reyes et al. 2000). In addition, as mentioned previously, a small percentage of TMAO is found in the urine following exposure to DMA. The formation of TMAO would require the reduction of DMA^V to DMA^{III} before the addition of the third methyl group, indicating DMA^V reductase activity.

The activities of the arsenic reductases appear to vary markedly among tissues. In the male hamster, the highest activity was found in the brain, followed by the bladder, spleen, and liver; the lowest activity was found in the testis (Sampayo-Reyes et al. 2000). Therefore, the tissues with high arsenic reductase activity seem to be different from those with high arsenic methyltransferase activity (testis > kidney > liver > lung, as discussed previously). Considering the marked species differences in the metabolism of arsenic, more data are needed on the tissue variation in arsenic metabolizing enzymes in various species. In particular, more data are needed on tissue variations in enzyme activities in humans. More data are also needed on sex differences, because most experiments have been performed in male animals.

There is increasing evidence that the trivalent methylated arsenic metabolites (especially MMA^{III}) are released from the site of arsenic methylation. Aposhian et al. (2000a) reported that people in Romania exposed to arsenic in drinking water (28, 84, or 161 µg/L) had MMA^{III} in their urine at concentrations of 5 -7 µg/L, irrespective of their exposure. Mandal et al. (2001) reported the presence of both MMA^{III} (2-5% of urinary arsenic) and DMA^{III} (5-20% of urinary arsenic) in the urine of subjects chronically exposed to inorganic arsenic via drinking water (33-250 µg/L) in four villages in West Bengal, India. The concentrations of MMA^{III} and DMA^{III} in the four villages ranged from 3-30 µg/L and 8-64 µg/L, respectively. The concentrations of both MMA^{III} and DMA^{III} increased with increasing concentration of total arsenic in urine (i.e., the sum of metabolites). It should be noted that the concentrations of trivalent metabolites in the urine might have been underestimated, because the trivalent metabolites are easily oxidized (Le et al. 2000). On the other hand, the trivalent metabolites are reported to be highly reactive; therefore, it is unlikely that urinary concentrations of the trivalent metabolites would be as high as MMA^{V} and DMA^{V}, which are much less reactive and readily excreted in urine.

There are very few data on the tissue distribution of trivalent methylated arsenic metabolites following exposure to inorganic arsenic, and no data in humans. Bile-duct-canulated rats were injected intravenously with arsenite or arsenate. Almost 10% of the injected dose (50 µmol/kg) was excreted in the bile as MMA^{III} and As^{III} (Gregus et al. 2000). Rats excrete much more arsenic in the bile than other species; therefore, it is difficult to generalize those results to other species. However, further support for the formation of trivalent methylated arsenic compounds in vivo following exposure to inorganic arsenic comes from experiments in hamsters. Both MMA^{III} and DMA^{III} were detected in the livers of hamsters treated with arsenite (Sampayo-Reyes et al. 2000).

Some of the trivalent methylated arsenic metabolites found in urine might be, in part, the result of reduction occurring in the kidneys and urinary bladder. It has been shown that arsenate is reabsorbed and reduced in the proximal renal tubuli, after which As^{III} is excreted in the urine (NRC 1999). Also, as mentioned above, high MMA^V reductase activity was found in the bladder of the hamster (Sampayo-Reyes et al. 2000), indicating that reduction of MMA, and possibly also DMA, might occur in the urinary bladder.

Administration of 300 mg of the chelating agent sodium 2,3-dimercapto-1-propane sulfonate (DMPS) to people exposed to arsenic in drinking water (568 ± 58 µg/L) in Inner Mongolia, China, markedly increased the urinary concentrations of inorganic arsenic (50-125 µg/L, on average) and MMA (50-325 µg/L), while the concentration of DMA decreased (240-125 µg/L) (Aposhian et al. 2000a; Le et al. 2000). This study was the first to identify MMA^{III} in urine as one of the arsenic metabolites. In vitro studies using partially purified rabbit liver MMA^{III} methyltransferase showed that the MMA^{III}-DMPS complex did not serve as a substrate for the enzyme (Aposhian et al. 2000a). Therefore, the authors suggested that DMPS forms a stable complex with MMA^{III}, which is excreted in urine.

It should be noted that the amount of MMA^{III} formed in tissues following exposure to inorganic arsenic is dependent upon the activity of arsenite methyltransferase, the enzyme that forms MMA^V, and the presence of MMA^V reductase, the enzyme that reduces MMA^V to MMA^{III}. It is also dependent on the presence and activity of MMA^{III} methyltransferase, the enzyme that further methylates MMA^{III} to DMA^V, which is readily excreted from cells (Styblo et al. 2000).

Inorganic As^{III} and the reduced forms of the methylated arsenic metabolites (MMA^{III} and DMA^{III}) are highly reactive and might contribute to the toxicity observed following exposure to inorganic arsenic (see Mechanisms of Toxicity). As discussed in the previous NRC report (NRC 1999) and by Styblo et al. (1997), trivalent inorganic and methylated arsenic metabolites have been shown to complex with GSH. Also, trivalent arsenicals are believed to form highly stable complexes with molecules containing vicinal thiols. In studies investigating the binding of As^{III} to proteins following exposure to arsenite in human lymphoblastoid cells, at least four 20-50 kDa proteins with arsenic affinity were isolated (Menzel et al. 1999). Two of the proteins identified were tubulin and actin. This ability of arsenic to bind to functional groups (e.g., thiols) can result in the inhibition of certain enzymes (Lin et al. 2001) and is one possible mechanism underlying arsenic's toxicity. Reactions with sulfhydryl groups are also the basis for arsenic detoxification therapy (e.g., use of DMPS) (Aposhian et al. 2000a).

ANIMAL TOXICITY STUDIES

Animal Bioassays

Although arsenic is not typically positive in experimental animal carcinogenicity bioassays (NRC 1999), several recent experiments have demonstrated that, under certain conditions, some forms of arsenic can induce tumors in animals.

For a recent doctoral dissertation, which was also published as a conference proceeding, Ng (1999) administered As^V at 500 µg/L in drinking water to 90 female C57BL/6J and 140 female metallothionein (MT) knock-out transgenic mice for 26 months. Control groups of 60 mice received tap water containing arsenic at less than 0.1 µg/L. Weekly water consumption and weight gain were reported, and survival was high (i.e., survival rates of 81% for C57BL/6J exposed mice, 74% for MT knock-out exposed mice, 98% for C57BL/6J control mice, and 97% for MT knock-out control mice after 2 years). The data indicate that the exposed mice developed more tumors than the controls by study termination. The subcommittee's understanding is that the results of this study are undergoing further examination and should be regarded as preliminary. If final analyses confirm the preliminary report, those results will be important because they would be the first data to demonstrate tumors in animals following ingestion of inorganic arsenic.

Waalkes et al. (2000) injected rats intravenously (i.v.) with sodium arsenate at 0.5 mg/kg once weekly for 20 weeks; the study was completed after 96 weeks. Significant skin changes (hyperkeratotic lesions) and renal lesions occurred in arsenate-treated females. Although these repeated arsenate exposures were not tumorigenic outright, there was clear evidence of proliferative, preneoplastic lesions of the uterus, testes, and liver. The authors suggested that because estrogen treatment has been associated with proliferative lesions and tumors of the uterus, female liver, and testes in other studies, arsenate might somehow act through an estrogenic mode of action. That hypothesis is supported by the observation that arsenate-induced uterine hyperplastic lesions showed a strong up-regulation of cyclin D1, an estrogen-associated gene product, and an up-regulation of estrogen receptor (ER) immunoreactive protein in the early lesions of uterine luminal and glandular hyperplasia.

Simeonova et al. (2000) administered sodium arsenite in drinking water to mice at doses of either 0, 0.002%, or 0.01% (20,000 or 100,000 ppb) for up to 16 weeks. After 4 weeks of exposure to the high dose (100,000 ppb), mouse urothelium exhibited hyperplasia, accompanied by accumulation of inorganic trivalent arsenic and, to a lesser extent, DMA. The authors found a persistent increase in DNA binding of the nuclear transcription factor, AP-1, in bladder epithelium of arsenite-exposed mice at both doses. Using a transgenic strain of mice possessing a luciferase reporter gene containing an AP-1 activation site, they were able to demonstrate that arsenite induces AP-1-mediated transcriptional activation in the urothelium in vivo. Using several tools to determine alterations in gene transcription, including cDNA expression arrays, the authors also demonstrated that arsenite alters the expression of a number of genes associated with cell growth, including the oncogenes c-*fos*, c-*jun,* and *EGR-1*. The expression of genes involved in cell-cycle arrest, including *GADD153* and *GADD45*, were enhanced by arsenite. The authors concluded that "the proliferation-enhancing effect of arsenic on uroepithelial cells likely contributes to its ability to cause cancer." In a follow-up study (Simeonova et al. 2001) in which groups of mice were exposed to a range of arsenite concentrations in drinking water, an increase in AP-1 DNA binding in bladder epithelial tissue was detected after 8 weeks of exposure to drinking water containing 50,000 µg/L and 100,000 µg/L, but not after exposure to water containing arsenite at 500 µg/L and 20,000 µg/L.

Santra et al. (2000) examined the hepatic effects of chronic ingestion (up to 15 months) of drinking water containing arsenic (1:1 arsenite to arsenate) at 3.2 mg/L (3,200 µg/L) in male BALB/c mice. Groups of arsenic-exposed mice and unexposed controls were sacrificed at 3, 6, 9, 12, and 15 months for examination of hepatic histology and certain biochemical parameters of oxidative stress. Statistically significant decrements in body weight appeared in the exposed animals at 12 months and 15 months, without significant differences between exposed and control groups in the amount of food or water consumption. No abnormal hepatic morphology was observed by light microscopy during the first 9 months of arsenic exposure, but at 12 months, 11 of 14 mice in the experimental group exhibited hepatocellular degeneration and focal mononuclear cell collection. After 15 months, exposed mice displayed evidence of hepatocellular necrosis, intralobular mononuclear cell infiltration, Kupffer cell proliferation, and portal fibrosis. Hepatic morphology was normal in all control mice. Biochemical changes consistent with oxidative stress preceded the overt histological pathology: hepatic glutathione was significantly reduced after 6 months, hepatic catalase was significantly reduced at 9 months, and hepatic glutathione *S*-transferase and glutathione reductase activities were significantly reduced at 12 and 15 months. There was a progressive, time-dependent increase in lipid peroxidation, as evidenced by increased production of malondialdehyde, and concomitant time-dependent damage to hepatocellular plasma membranes, as evidenced by decreases in membrane Na^+/K^+ ATPase activity. The findings of Santra et al. (2000) represent the first animal model to demonstrate hepatic fibrosis following chronic arsenic ingestion. Periportal fibrosis, sometimes with noncirrhotic portal hypertension, is a recognized sequela of chronic arsenic ingestion in humans (Santra et al. 1999; NRC 1999). Biochemical changes observed in this long-term in vivo animal-feeding experiment suggest that these adverse effects of arsenic may be mediated through oxidative stress.

In the past, DMA^V has been considered primarily a detoxification product in arsenic metabolism that is rapidly eliminated from cells and excreted in the urine. However, over the past several years, there has been mounting evidence that DMA might possess biological activity that is relevant to arsenic carcinogenesis. Wei et al. (1999) found that chronic administration of DMA in drinking water, albeit at very high doses, to male Fischer 344 (F344) rats induced bladder cancer. The rats were given DMA at 0, 12.5, 50, or 200 parts per million (ppm) (0, 12,500, 50,000, 200,000 µg/L) in drinking water for 104 weeks. From weeks 97 to 104, urinary bladder tumors were seen in 8 of 31 rats (26%) at 50 ppm, and 12 of 31 rats (39%) at 200 ppm. No bladder tumors were seen in any of the control animals (36 rats) or in the 12.5-ppm group (33 rats).

In a study of lung tumors in mice, Hayashi et al. (1998) found that administration of DMA at 400 ppm (400,000 µg/L) in drinking water for 50 weeks to A/J mice produced an increase in the number of pulmonary tumors per mouse (1.36 and 0.5 tumors per mouse in exposed and control groups, respectively), and the authors concluded that DMA alone can act as a carcinogen in mice.

Neither of the studies by Wei et al. (1999) and Hayashi et al. (1998) provides dose-response data, but the types of tumors observed (bladder and lung tumors) are highly relevant to what has been seen in human epidemiological studies.

Yamanaka et al. (2000) administered DMA at either 400 or 1,000 ppm (400,000 or 1,000,000 µg/L) in drinking water to hairless mice that were concomitantly exposed to ultraviolet B radiation (2 kilojoules per square meter (kJ/m^2) twice weekly) for 25 weeks. The number of malignant skin tumors per mouse was slightly, but significantly, increased in the 1,000-ppm treatment group, but no differences from the control were seen in the 400-ppm group. There were no differences in the percentage of tumor-bearing mice over the 25-week period. Therefore, in this study, there was only a modest cocarcinogenic effect seen at a very high dose of exogenously administered DMA.

In addition to these bioassays of DMA, another study found that administration of DMA in drinking water (100 ppm, or 100,000 µg/L, for 36 weeks) can act as a bladder tumor promoter in Lewis x F344 strain of rats initiated with nitrosamine in drinking water (Chen et al. 1999). Cohen et al. (2001) showed that DMA at 100 ppm (100,000 µg/L) in the diet of female F344 rats produced cytotoxic changes in the rat urothelium as early as 6 hr after exposure was begun. Necrosis was evident by 3 days, followed by regenerative hyperplasia of the bladder epithelium. Morikawa et al. (2000) found that 3.6 mg of DMA applied topically to a skin-tumor sensitive strain of mice twice a week for 18 weeks significantly accelerated skin tumor development. The authors concluded that DMA has a promoting effect on skin tumorigenesis in this strain of mice.

The cancer bioassays of DMA utilized extremely high doses administered to rats. Their interpretation is limited by the lack of any studies relating the relative intracellular concentrations of DMA achieved by direct administration of DMA in drinking water to the intracellular concentrations derived from biomethylation of arsenic to DMA following chronic exposure to inorganic arsenic in drinking water. The relative uptake of exogenously administered DMA into cells has not been determined directly, but Males et al. (1998) found that the transmembrane permeability coefficient for DMA was two orders of magnitude greater than that for MMA when measured in unilamellar vesicles, suggesting that DMA may readily diffuse across cell membranes. Furthermore, trimethylarsine oxide has been identified in human urine following administration of DMA[V], indicating that some uptake and metabolism of DMA does occur in humans (for review, see Kenyon and Hughes 2001).

In summary, several new animal bioassays of various forms of arsenic have been reported since the previous NRC report (NRC 1999). Although these studies have demonstrated "positive" carcinogenic responses to arsenic in rodent species under certain conditions, the doses used were very high relative to human exposures to arsenic via drinking water. Relatively few doses were used, and information on the shape of the in vivo dose-response curve is limited at best. Although these studies are of qualitative interest in demonstrating the carcinogenic potential of arsenic in rodents, the studies provide little useful quantitative information relevant to human risk assessment of arsenic following exposure via drinking water and cannot replace the current body of epidemiological data used for risk assessment.

Species Differences in Toxicity

As noted in the 1999 NRC report, there are differences in susceptibility to arsenic among animals. As discussed earlier in the Toxicokinetics section, species differences have also been noted in more recent studies. Mice were found to be less susceptible than rats to DMA toxicity (Ahmad et al. 1999; Brown et al. 1997). Mice dosed orally with DMA at 720 mg/kg showed no DNA damage in the lung, no induction of cytochrome P-450 in the liver, and no reduction in serum alanine aminotransferase activity. In contrast, all of those changes were observed in rats dosed with DMA at 387 mg/kg. Both species had reduced liver GSH and reduced lung ornithine decarboxylase activity at the doses administered.

Rats and guinea pigs, but not mice, showed an accumulation of copper in the kidneys following administration of sodium arsenite at levels of 0, 10, 30 or 60 mg/kg of diet for 1, 2, or 3 weeks (Hunder et al. 1999). The authors attributed the species difference to more efficient methylation and elimination of arsenic by mice than by rats and guinea pigs. The copper content of the renal cortex increased from 10 µg/g of wet weight to as high as 65 µg/g of wet weight in a time- and concentration-dependent manner.

Developmental Toxicity Studies

One recent animal study has been published on the developmental toxicity resulting from repeated oral dosing with arsenic compounds. Data on the developmental effects of arsenic in humans is discussed in Chapter 2. Arsenic trioxide was repeatedly administered orally to Crl:CDO(SD)BR rats at doses of 0, 1, 2.5, 5, or 10 mg/kg/day (Holson et al. 2000). No evidence of developmental toxicity was found at doses that did not cause maternal toxicity.

Another study examined the effect of genotype in mice on arsenite-induced congenital malformations in the neural tube (Machado et al. 1999). Mice that were bred to include the "splotch" allele were more sensitive to the teratogenic effects of arsenite, illustrating that genotype can influence arsenic toxicity. It should be noted that doses of arsenicals required to elicit teratogenesis acutely in outbred strains of mice are orders of magnitude higher than doses to which humans are exposed environmentally.

MECHANISMS OF TOXICITY

Numerous studies have been completed in the past 3 years that address the potential modes of action and specific mechanisms by which arsenic exerts its toxic effects, including cancer. Four major, and overlapping, areas have received much attention: (1) induction of mutations and chromosomal abberations; (2) alterations in signal transduction, cell-cycle control, differentiation, and apoptosis; (3) induction of oxidative stress; and (4) alterations in gene expression. It is important to recognize that these categories of potential modes of action are not distinct, and the contribution of a specific mechanism (e.g., particular type of mutation, inhibition of a specific enzyme or pathway, or induction of a specific form of reactive oxygen species) within each category is difficult to determine. The sensitivity of those targets would be expected to be dose- and time-related and might also be species and tissue specific.

Because various forms of arsenic have different toxic potencies or modes of action, some discussion of the new findings on the relative differences is warranted before discussion of the new data on each potential mode of action listed above. Resistance and tolerance to the cytotoxicity of arsenic and the implications of the experimental data on the carcinogenicity of arsenic are also discussed.

Table 3-1 provides an overview of some of the mechanistic studies that have been completed since 1998. The subcommittee focused on studies that appear to induce biochemical effects at moderate to relatively low concentrations of arsenic in vitro (e.g., less than 10 μM), although a few studies that used higher concentrations are included for comparative purposes. Studies that require arsenic concentrations greater than 10 μM to produce a biological response in vitro are less likely to be relevant to the health effects related to chronic ingestion of arsenic in drinking water and have not been exhaustively reviewed.

Relative Toxicity of Different Forms of Arsenic

In the previous NRC report, it was stated that the pentavalent methylated arsenic metabolites are much less mutagenic than the inorganic arsenicals (NRC 1999; p.199). At that time, few studies had directly assessed the relative toxicity and mutagenicity of the trivalent methylated forms of arsenic.

TABLE 3-1 Overview of Recent In Vitro Studies Showing Arsenic Biological Activity in Low Micromolar to Submicromolar Concentration Range

Model System	Arsenic Species	Concentrations Tested (μM)	LOEC (μM) [a]	End Points Measured[b]	Reference
Cytotoxicity					
Human cultured keratinocytes (SV40 transformed)	As^{III}	0.1-3	1.1 (IC_{50})	Increased neutral red uptake	Snow et al. 1999
Chang human liver cells	$MMAs^{III}$ As^{III}	1.25-20 20-500	5 20	Increased LDH and K^+ release	Petrick et al. 2000
Cell Proliferation Stimulation					
Human epidermal keartinocytes	As^{III}	0.001-0.01	0.001	Stimulation of cell proliferation	Vega et al. 2001
Human keratinocytes (HaCaT)	As^{III}	0.5-1.0	0.5	Increased growth rate	Chiang et al. 2001
Cell Cycle Arrest and Apoptosis					
Myeloma cell lines	As^{III}	10 nM-1	0.3	Apoptosis	Rousselot et al. 1999
Human gastric carcinoma cells	As^{III}	0.01-1	0.01	Induction of apoptosis	Zhang et al. 1999
K562 (erythroid myeloid) U937 (macrophage-like) HL60 (promyelocytic) cells	As^{III}	0.5-5	2.5	Cell-cycle arrest in metaphase	Li and Broome 1999

Primary rat cerebellar neurons	As^{III}	5-15	5	Apoptosis	Namgung and Xia 2001
Human cancer cells: Bladder (5 lines) Lung (2 lines) Liver (2 lines) Others (10 lines)	As^{III}		IC_{50}s: 0.7-5.2 6.6, 8.4 10.4, 14.4 1.3-6.8	Growth inhibition/survival after 96 hr	Yang et al. 1999
Human myeloma cells	As^{III}		1-2 (IC_{50})	Cell-cycle arrest; apoptosis	Park et al. 2000
PCCL-1 cells	As^{III}		2	Cell-cycle arrest; apoptosis	Seol et al. 1999
Myeloma cell lines	As^{III}	0.5-1	0.5	Cell-cycle arrest; apoptosis	Deaglio et al. 2001
Inhibition of Energy Cycle					
Cell-free system (purified porcine enzyme)	As^{III} $MMAs^{III}$	0.5-1,000	106 18	Inhibition of pyruvate dehydrogenase	Petrick et al. 2001
Human cultured keratinocytes (SV40 transformed)	As^{III}	Not stated	5.6 (IC_{50})	Inhibition of pyruvate dehydrogenase activity	Snow et al. 1999
Cell-free system (enzyme source unspecified)	As^{III}		5.6	Inhibition of pyruvate dehydrogenase	Hu et al. 1998
Effects related to Oxidative Stress					
Human cultured keratinocytes (SV40 transformed)	As^{III}	0.05-5	0.05-0.5	Interaction with H_2O_2 in neutral red-dye assay	Snow et al. 2001
Rat cultured hepatocytes	$MMAs^{III}$	0.1-10	1	Decreased thioredoxin reductase activity	Lin et al. 2001

Continued

TABLE 3-1 *Continued*

Model System	Arsenic Species	Concentrations Tested (µM)	LOEC (µM)[a]	End Points Measured[b]	Reference
Purified thioredoxin enzyme from mouse liver	MMAs[III]	0.2-1,000	<0.5 (IC$_{50}$=0.7)	Inhibition of thioredoxin reductase	Lin et al. 1999
Purified thioredoxin enzyme from mouse liver	As[III]	<1-1,000	100	Inhibition of thioredoxin reductase	Lin et al. 1999
Rat cultured hepatocytes	As[III]	1-50	50	Decreased thioredoxin reductase activity	Lin et al. 2001
Human cultured keratinocytes (SV40 transformed)	As[III]	3	3	Acute up-regulation of reduced glutathione (GSH) and GSH-related enzymes	Snow et al. 1999
Human-hamster hybrid cells	As[III]	30 (4 µg/mL)	30	Production of ROS, measured by electron spin resonance (ESR)	Liu, S.X. et al. 2001
Human cultured keratinocytes (SV40 transformed)	As[III]	Not stated	2.0 (IC$_{50}$)	Inhibition of glutathione peroxidase activity	Snow et al. 1999
Interference with Hormone Function					
Hormone responsive H4IIE rat hepatoma cells	As[III]	0.3-3.3	0.3	Function of the glucocorticoid receptor	Kaltreider et al. 2001
DNA Damage and Repair					
Human hepatocyte cell line WRL-68	As[III]	0.001-1	0.001	Induction of protein-DNA cross links	Ramirez et al. 2000

Human kidney carcinoma cell lines		0.002–0.2	0.009	Colony-forming efficiency; altered DNA methylation	Zhong and Mass 2001
Human cultured keratinocytes (SV40 transformed)	As^{III}	0.1–25	0.1–0.5	Acute up-regulation of DNA repair genes, thioredoxin, glutathione reductase; decreased expression of glutathione peroxidase	Snow et al. 2001
Mouse C3H 10T1/2 preadipocytes	As^{III}	0.1–50	3	Inhibition of dexamethasone/insulin-induced lipid accumulation; accentuated response to mitogenic stimulation	Trouba et al. 2000a
Human vascular smooth muscle cells	As^{III}	1–10	1	Comet assay for DNA damage, increased NADH oxidase	Lynn et al. 2000
Human fibroblasts	As^{III}	1.25–10	14.5 (IC_{50})	Micronuclei	Yih and Lee 1999
Human cultured keratinocytes (SV40 transformed)	As^{III}	10–100		Inhibition of DNA ligase activity and DNA repair	Snow et al. 1999
Rat liver cells, TRL1215	As^{III}		0.125	Depletion of S-adenosylmethionine (SAM); DNA hypomethylation; altered gene expression (80/588)	Chen et al. 2001

Continued

TABLE 3-1 *Continued*

Model System	Arsenic Species	Concentrations Tested (µM)	LOEC (µM)[a]	End Points Measured[b]	Reference
Human lymphocytes (PHA-stimulated)	As[III]		10	Oxidative damage to DNA (measured by Comet assay)	Li et al. 2001
Gene Expression					
Human cultured keratinocytes	As[III]	0.001-1	0.001	Decreased p53 protein; increased mdm2 protein	Hamadeh et al. 1999
Mouse hepatoma cells (Hepa-1)	As[III]		??	Enhanced TCDD-inducible levels of Nqo1 mRNA expression	Maier et al. 2000
Rabbit renal cortical slices	As[III]	0.01-10		Induction of HO-1 and ATF-2, enhanced DNA binding of AP-1, and Elk-1	
Rabbit renal cortical slices	As[V]	0.01-10		Induction of ATF-2, enhanced DNA binding of AP-1, and Elk-1	
Mouse epidermal cell line	As[III]		0.8	Activation of ERK phosphorylation	Huang et al. 1999a
W138 human fibroblast	As[III]		0.1	Increased p53 protein, and cyclin D1	Vogt and Rossman, 2001

Gene Mutation

Human-hamster hybrid cells (A$_L$; contains human chromosome 11)	As^{III}	(0.1-2 µg/mL) 3.8	Increased mutation rate in CD59 gene	Liu, S.X. et al. 2001

[a] The LOEC represents the lowest concentration in the cited study at which an effect was noted. Because of variations in study design, it should not necessarily be concluded that effects could not occur at a lower concentrations.

[b] The reported end points occurred after exposure at sublethal doses.

Abbreviation: LOEC, lowest observed effect concentration.

Aposhian and colleagues (Petrick et al. 2001) recently evaluated the relative toxicities of MMA^{III} and As^{III}, in the hamster. Single LD$_{50}$s (lethal dose to 50% of animals) in hamsters were calculated to be 29 and 113 micromoles (μmol)/kg for MMA^{III} and As^{III}, respectively. Those data represent the first in vivo acute toxicity data for MMA^{III} and show, somewhat surprisingly, that it is about 4 times more acutely toxic than inorganic arsenite. Earlier LD$_{50}$ studies showed that the pentavalent species MMA^{V} is less toxic than arsenite by more than one order of magnitude in some species (for a tabular comparison, see ATSDR 2000). Petrick et al. (2001) also found that MMA^{III} was a much more potent inhibitor of hamster-kidney and porcine-heart pyruvate dehydrogenase (PDH) than inorganic arsenite. Previous studies demonstrated that As^{III} was a potent inhibitor of PDH via covalent binding of the arsenite to critical vicinal dithiols on the enzyme. Because MMA^{III} is in the same oxidative state as As^{III}, the authors suggested that the mechanism for MMA^{III} is likely to be the same. Using nuclear magnetic resonance (NMR) spectroscopy, they also showed that methylarsine oxide (MMA^{III} oxide, $(CH_3AsO)_n$) and diiodomethylarsine (MMA^{III} iodide, CH_3AsI_2), the chemical forms of MMA^{III} administered to the animals and used in the in vitro studies, hydrolyze to MMA^{III} ($CH_3As(OH)_2$) in aqueous solution.

Several other new studies investigated the toxicity of different forms of arsenic and its methylated derivatives in cultured cells. Moderate amounts of As^{III} inhibited the synthesis of DMA, resulting in an accumulation of As^{III} and monomethyl arsenic species in the cells. The same investigators (Petrick et al. 2001) found that, in vivo, MMA was more toxic than arsenite when given to hamsters intraperitoneally (LD$_{50}$ = 29.3 μmol/kg and 112 μmol/kg, respectively). Using primary cells from human and rat liver, human neonatal foreskin, and human cervix as well as a simian virus (SV)-40-transformed epithelial cell line derived from normal human urinary bladder cells, the investigators found that the trivalent methylated metabolites of arsenic were more cytotoxic than arsenite in all cell types (Styblo et al. 1999b). The same group further reported that MMA^{III} was the most cytotoxic in all cell types (including primary human bronchial epithelial cells) and that DMA^{III} was at least as toxic as As^{III} (Styblo et al. 2000). Pentavalent arsenicals were much less cytotoxic than the trivalent forms in all cell lines. The transformed cell line derived from human bladder cells had no capacity to methylate the various forms of arsenic. It is uncertain whether primary epithelial cells lack the methylation capacity or whether the transformation process causes the lack of capacity. There was no apparent correlation, however, between susceptibility of cells to arsenic toxicity and their capacity to methylate As^{III}. Similar findings were reported by Petrick et al. (2000), who tested the relative toxicities of different forms of arsenic in Chang human hepatocytes and found that the order of toxicity was monomethylarsonous acid > arsenite > arsenate > monomethylarsonic acid = dimethyl arsinic acid. A major conclusion of all of these studies is that, contrary to what was thought earlier, methylation of arsenicals is not entirely a detoxification process, and cells with a high capacity to methylate arsenic compounds are not necessarily protected from arsenic toxicity.

Induction of Mutations and Chromosomal Abberations

The 1999 NRC report provided a detailed review of many studies documenting that arsenic is not a particularly effective direct-acting mutagen, but it is quite effective in altering chromosomal integrity. Several recent in vitro and in vivo studies have examined the impact of arsenic on chromosomal aberrations and gene mutations.

Mass et al. (2001) utilized a DNA-nicking assay of phage DNA and the single-cell "comet" assay in cultured human lymphocytes to assess the genotoxic properties of inorganic and various methylated arsenicals. The nicking assay examines the ability of arsenic compounds to induce breaks in naked DNA in a cell-free system without added enzymes or chemical-activation systems. MMA^{III} and DMA^{III} were the only forms of arsenic able to directly damage naked DNA in this assay. As noted by the authors, MMA^{III} was effective at nicking DNA at 30 mM; however, at 150 µM of DMA^{III}, nicking could be observed. Although the mechanism of the DNA damage was not determined, the authors referred to preliminary studies in which the nicking activity was not inhibited by the addition of glutathione or by the use of argon purge (i.e., under reduced oxygen conditions), suggesting that oxidative stress might not have been responsible. In the comet assay, which examined the ability of arsenicals to induce DNA damage in cultured human lymphocytes, MMA^{III} and DMA^{III} were 77 and 386 times more potent, respectively, than arsenite. The relative potency was determined from the slope of the dose-response curves for MMA^{III} (from 1.25 to 80 µM), DMA^{III} (from 1.4 to 91 µM), and As^{III} and As^{V} (from 1 to 1,000 µM). The study by Mass et al. (2001) is noteworthy because it provides additional evidence that methylation of arsenic may not be a detoxification pathway and suggests that the trivalent methylated arsenicals might be capable of causing genotoxic damage by acting directly on DNA.

Human fibroblasts exposed to arsenite from 1.25 to 10 µM for 24 hr induced micro nuclei formation in human fibroblasts in culture. Micro nuclei formation was blocked by both catalase and *N*-acetyl cysteine, suggesting that the induction of micro nuclei was mediated via arsenic-induced oxidative stress (Yih and Lee 1999). As noted in the 1999 NRC report, several in vivo studies have associated arsenic exposure with the appearance of micronuclei in human cells, including a report of micronuclei in the exfoliated bladder cells of men in Chile who consumed drinking water with an arsenic concentration of 54 to 137 µg/L (Moore et al. 1997; Biggs et al. 1997).

Liou et al. (1999) investigated the utility of chromosomal aberrations in lymphocytes as a biomarker of cancer risk. In a nested, case-control study of cancer incidence in a cohort of 686 subjects from the area of southwestern Taiwan where blackfoot-disease is endemic, 31 cases of cancer developed during a 4-year follow-up (11 skin, 4 bladder, 3 lung, and 3 uterine and cervical cancers). Lymphocyte cytogenetics could be examined in 22 cases and 22 matched controls for whom the duration of consumption of artesian well water was approximately 30 years. No significant differences were seen in the frequencies of sister-chromatid-exchange (SCE) and chromatid-type aberrations in the cases and the controls. The frequency of chromosomal-type aberrations, such as gaps, breaks, and breaks plus exchanges, and the total frequency of chromosomal-type aberrations were significantly higher ($p < 0.05$) in cases than in controls. The odds ratio for cancer risk in subjects with at least one chromosomal-type break was 5.0 (95% confidence interval = 1.09 to 22.82). The odds ratio for cancer risk in subjects with at least one chromosomal break plus exchanges and a frequency of total chromosomal-type aberrations greater than 1.007% are 11.0 and 12.0, respectively ($p < 0.05$). Subjects with a frequency of total chromosomal-type aberrations greater than 4.023% had a 9-fold increase in cancer risk. Those results suggest that chromosomal-type aberrations in lymphocytes might be useful biomarkers for the prediction of cancer development in arsenic-exposed populations.

In a small study of subjects exposed to arsenic in drinking water in Finland, Maki-Paakkanen et al. (1998) examined the association between urinary arsenic concentration and chromosomal aberrations (gaps, isogaps, breaks, and rearrangements) in human lymphocytes. The total urinary arsenic concentration (sum of inorganic arsenic, MMA, and DMA) was 180 µg/L (from 7 to 500 µg/L) in 27 exposed subjects who were consuming well water, 17 µg/L (from 3 to 104 µg/L) in nine individuals who had previously consumed the water, and 7 µg/L (from 4 to 44 µg/L) in eight controls. In multiple regression models adjusting for age, gender, and smoking, total urinary arsenic was positively associated with the number of chromosomal aberrations ($r^2 = 0.23$, $p = 0.02$ when chromosomal gaps were excluded; $r^2 = 0.15$, $p = 0.07$ when chromosomal gaps were included). Among current users of the well water, the ratio of MMA-to-total arsenic in urine was positively correlated with the number of chromosomal aberrations (including gaps), and the ratio of DMA-to-total arsenic in urine was negatively correlated with the number of chromosomal aberrations (including gaps).

One recent study showed that As^{III} can induce DNA-protein cross-links (DPCs) in vitro in a human liver cell line at concentrations as low as 1 nanomolar (nM) (Ramirez et al. 2000). The extent of DNA-protein cross-links increased linearly with arsenite concentration. Detectable DPCs decreased within a few hours of substituting arsenite with an arsenite-free culture medium, possibly indicating their elimination through DNA repair. Presumably, because of their high cysteine (and hence thiol) content, several cytokeratins were identified by immunoblotting techniques to be among the proteins that cross-linked with DNA. The total cellular content of cytokeratin CK18 was increased following arsenite exposure. Ramirez et al. (2000) suggested that the formation of arsenite-induced DPCs might be involved in arsenite-related chromosomal aberrations. They also suggested that arsenite's effect on cytokeratins might interfere with the important role of these proteins in cellular differentiation.

Alterations in Signal Transduction, Cell-Cycle Control, Differentiation, and Apoptosis

Although in vitro studies can provide useful information on putative mechanisms of arsenic toxicity and carcinogenicity, it is important to recognize the many limitations associated with such studies. Quantitative and qualitative responses of an in vitro system to arsenic might be influenced by factors such as species of arsenic used, dose and time frame of exposure, period and nature of observations, experimental and culture conditions (e.g., state of confluency, presence or absence of serum, nature and concentration of other nutrient supplements), and cell type and tissue origin. As discussed in Chapters 2 and 4, there are species differences in response to the toxic and carcinogenic actions of arsenic, and it is also likely that there is interindividual variability in response. Mechanistic studies are helpful in identifying potential reasons for such variability in responses observed in vivo, but great caution should be used in inferring relevance of a specific response to arsenic observed in vitro to humans exposed chronically at relatively low doses in drinking water.

Cell Signaling

Trouba et al. (2000b) found that long-term exposure of epithelial-growth-factor (EGF)-stimulated murine fibroblasts to arsenite increased DNA synthesis and the proportion of cells entering S phase. Expression of the positive growth regulators c-myc and E2F-1 were both increased, although extracellular signal-regulated protein kinases (ERK-2s) and EGF-receptor expression were unchanged. The negative regulators of proliferation, mitogen-activated protein (MAP) kinase phosphatase-1 and p27 (Kip1), were lower in As^{III}-treated cells compared with control cells. The authors concluded that long-term exposure to high levels of arsenite might make cells more susceptible to mitogenic stimulation and that alterations in mitogenic signaling proteins might contribute to the carcinogenic actions of arsenite.

Porter et al. (1999) evaluated the role of different mitogen-activated protein kinases/extracellular signal-regulated kinase kinase (MEKK) kinases that are involved in arsenate and arsenite-induced activation of c-Jun N-terminal kinases (JNKs). Interestingly, both arsenate and arsenite activated JNK but, apparently, by slightly different pathways. Arsenite-mediated activation of JNK requires MEKK2, MEKK3, and MEKK4, whereas arsenate required only MEKK3 and MEKK4. In addition, arsenite, but not arsenate, activation of JNK requires p21-activated kinase, whereas both forms of arsenic require the guanosine triphosphatases (GTPases) Rac and Rho.

Chen et al. (2000a) found that As^{III} induced the translocation of several protein kinase C (PKC) proteins from cytosol to membranes and activated activator protein-1 (AP-1). Using selective PKC inhibitors and dominant negative mutants, the authors demonstrated that PKC_δ, PKC_ε, and PKC_α mediate arsenic-induced AP-1 activation through different MAP kinase pathways (e.g., ERKs, JNKs, and p38 kinases).

Arsenite has been shown to inhibit the NF-κB signal transduction pathway that is important in transcriptional regulation of a variety of cellular pathways, including the inflammatory cytokines TNF-α and IL-8 (Roussel and Barchowsky 2000). Kapahi et al. (2000) identified what might be the first specific molecular target for arsenite. They found that arsenite is an effective inhibitor of I-κB kinase (IKK) (IC_{50}, 9.1 μM), which is required for the integrity and function of the NF-κB signaling pathway. As^{III} was shown to bind to Cys-179 in the activation loop of one of the IKK catalytic subunits. Replacement of the Cys-179 with alanine provides a fully functional IKK catalytic subunit but abbrogates As^{III} binding. Overexpression of the C179A mutant protects NF-κB from inhibition by arsenite, thereby demonstrating that direct binding of arsenite to the Cys-179 residue of IKK subunit is responsible for arsenite-induced inhibition of the NF-κB signaling pathway.

Vogt and Rossman (2001) examined the effect of arsenite on cell signaling in a cell culture of W138 normal human lung fibroblasts. Treatment of cells with 0.1 μM arsenite for 14 days caused a modest (3-fold) increase in the cellular levels of p53 protein without any increase in p21 protein, a major downstream protein involved in cell-cycle arrest that is often increased by signals associated with increases in p53. Arsenite exposure blunted the increase in p53 and prevented any increase in p21 when the cells were exposed to irradiation (6 Gy), an insult that alone markedly increased levels of both proteins. Arsenite at 0.1 μM for 14 days heightened expression of cyclin D1, a protein that facilitates progression through the G1 phase of the cell cycle. Vogt and Rossman (2001) suggested that the action of arsenite in preventing the p53-dependent p21-protein increase in cells that experience DNA damage, together with arsenite-induced up-regulation of cyclin D1, might promote replication of a DNA-damaged template. This action might be a mechanism for some of the observed comutagenic and cocarcinogenic properties of arsenite.

Collectively, the studies described above demonstrate that arsenic can interfere with cell signaling pathways (e.g., the p53 signaling pathway) that are frequently implicated in the promotion and progression of a variety of tumor types in experimental animal models and of some human tumors. However, which specific alterations in signal transduction pathways are actual targets that contribute to the development of arsenic-induced tumors in humans following chronic consumption of arsenic in drinking water remains uncertain.

Apoptosis, Cytotoxicity, and Anticancer Activity

Recent clinical trials have found that arsenite has therapeutic value in the treatment of acute promyelocytic leukemia, and there is interest in exploring its effectiveness in the treatment of a variety of other cancers (Soignet et al. 1999; Murgo 2001). It is of historical note that a therapeutic role for arsenic in the treatment of leukemia was suggested nearly a century ago (Osler 1894; Forkner and Scott 1931). A number of studies have examined the biochemical basis for this potential therapeutic effect (for reviews, see Chen et al. 2000b; Fenaux et al. 2001). Detailed descriptions of those studies are beyond the scope of this report; however, the potential mode of action of arsenic revealed from those studies might be relevant to the biological actions of arsenic in other human cell types.

In acute promyelocytic leukemia, the specific molecular event critical to the formation of malignant cells is known. Malignant lymphoblasts exhibit a t(15;17)-chromosomal translocation that produces a product that regulates expression of the tyrosine kinase oncogene BCR-ABL. Puccetti et al. (2000) demonstrated that 2 µM of As_2O_3 induces apoptosis in human lymphoblasts that carry the t(15;17) translocation, but not in wild-type cells, and also found that forced overexpression of BCR-ABL susceptibility in these cells resulted in greatly enhanced sensitivity to arsenic-induced apoptosis. From this observation, they concluded that As_2O_3 is a tumor-specific agent capable of inducing apoptosis selectively in acute promyelocytic leukemia cells.

Several recent studies showed that As_2O_3 can induce apoptosis through alterations in other cell signaling pathways. For example, Seol et al. (1999) found that 2 µM of As_2O_3 efficiently induced G2/M arrest in PCCI-1 cells, apparently in association with induction of p21 and reduction of cdc2 kinase activity. Numerous other studies found that low doses of As_2O_3 selectively induced apoptosis of NB4 promyelocytic leukemia cells (see review in Alemany and Levin 2000).

In addition to acute promyelocytic leukemia, As_2O_3 is thought to have therapeutic potential for myeloma. There are several recent studies on the ability of this agent to kill myeloma cells or T cells infected with leukemia virus. Studies on human myeloma-like cell lines indicated that the compound acted as an adjuvant in increasing cell killing by lymphocytes (Deaglio et al. 2001). Two cell-surface markers, CD38 and CD54m, were up-regulated, as were their corresponding ligands (CD31 and CD11a). Those data suggest that increased adhesion was responsible for the improved killing.

Park et al. (2000) examined the effect of As_2O_3 on proliferation, cell-cycle regulation, and apoptosis in human myeloma cell lines. They found that cell proliferation in eight different cell lines were inhibited by As_2O_3, with IC_{50}s (the concentration that inhibits the response by 50%) of 1-2 µM. Arsenic induced G1 and G2-M phase arrest. As_2O_3 markedly enhanced the binding of p21 to several cyclins and cyclin-dependent kinase 6. They concluded that As_2O_3 inhibits proliferation of myeloma tumor cells via cell-cycle arrest in association with induction of p21 and apoptosis. The apoptosis was associated with down-regulation of Bcl-2, loss of mitochondrial transmembrane potential, and an increase of caspase-3 activity. Earlier, Zhang et al. (1998) reported the induction of apoptosis and cell-cycle arrest in cell lines from lymphoid neoplasms by As_2O_3 at micromolar concentrations. Zhang et al.

(1999) also examined the sensitivity of six human cancer cell lines to As_2O_3-induced inhibition of cell growth and induction of apoptosis. They found that a gastric carcinoma cell line was particularly sensitive to As_2O_3-induced apoptosis, with concentrations as low as 0.01 µM causing significant growth inhibition and apoptosis.

Several studies reported that As_2O_3 induces cell-cycle arrest and apoptosis in malignant cells. Ishitsuka et al. (2000), using HTLV-1 infected T-cell lines, showed that As_2O_3 at concentrations that might be seen when it is used clinically inhibited cell growth by cell-cycle arrest and induced apoptosis. The induced apoptosis could be prevented by caspase inhibitors. They also observed destruction of the Bcl-2 protein and enhancement of the bak protein production. The arsenic compound increased expression of p53, Cip1/p21, and Kip1/p27 and dephosphorylation of retinoblastoma protein. Rousselot et al. (1999) found that As_2O_3 as well as the organic arsenical melarsoprol (2-[4-[(4,6-diamino-1,3,5-triazin-2-yl)amino]phenyl]-1,3,2-dithiarsolane-4-methanol) at pharmacological concentrations (10 nM to 1 µM) inhibited growth and induced apoptosis in myeloma cell lines. In bone-marrow samples, the compounds were able to induce apoptosis in myeloma cells while sparing the normal myeloid cells.

As_2O_3 induced apoptosis in PCI-1 head and neck cells after treatment for 3 days with 2 µM of arsenite (Seol et al. 2001). The mechanism appeared to be through up-regulation of Bcl-2; capase 9 was activated and the mitochondrial membranes of the cells were depolarized. To test the efficacy of the arsenic in vivo, C3H mice inoculated with syngenic SCC7 cells were treated by intratumoral injections of As_2O_3 (300 µg) daily for 4 days. The tumor was reduced in size and increased apoptosis was seen.

Other tumor cell lines have also been reported to be sensitive to arsenite trioxide. Uslu et al. (2000) reported the induction of apoptosis in prostate and ovarian carcinoma cell lines treated with 1 µM of As_2O_3. Jiang et al. (2001) reported the induction of apoptosis in human gastric cancer cells at concentrations as low as 0.1 µM. The treatment resulted in a marked increase in p53 protein levels and increased the activity of capase 3.

Shen et al. (1999) studied the concentration of arsenic trioxide required to inhibit growth and induce apoptosis in oesophageal cancer cells in vitro. The EC_{50} for induction of apoptosis was 1 µM.

In summary, numerous cancer chemotherapy studies in cell cultures and in patients with acute promyelocytic leukemia demonstrate that arsenic trioxide can lead to cell-cycle arrest and apoptosis in malignant cells.

A recent pilot study by Feng et al. (2001) examined DNA fragmentation by the TUNEL assay and the DNA ladder assay (indicators of apoptosis) in buccal epithelial cells from subjects consuming arsenic in drinking water in Inner Mongolia. Buccal cells were collected from 19 subjects exposed to a mean arsenic concentration in drinking water of 527.5 ± 23.7 µg/L and from 13 control subjects exposed to drinking water containing an arsenic concentration of 4.4 ± 1.0 µg/L. All the exposed subjects had arsenic-associated skin lesions. In the DNA ladder assay, 89% (17 of 19) of the arsenic-exposed group exhibited the positive finding of less than 100-base-pair (bp) fragmentation in comparison to 15% (2 of 13) of the controls ($p < 0.0001$). For the TUNEL assay, the mean frequency of positive staining cells was higher in the exposed group (15.1%) than in the control group (2.0%) ($p < 0.0001$). Differences between exposed and unexposed subjects remained significant after controlling for age, sex, and smoking. In nonparametric analysis, results from the DNA ladder assay and the TUNEL assay were highly correlated with arsenic concentration in drinking water ($r_{spearman} = 0.67$ and $r_{spearman} = 0.71$, respectively, both with $p < 0.0001$). There was indication of a synergistic effect between arsenic exposure and smoking on DNA fragmentation.

Exposure of primary cultures of rat cerebellar neurons to 5, 10, or 15 µM of arsenite induced apoptosis (Namgung and Xia 2001). DMA caused the same change but induced it at 1-5 µM concentrations. Co-treatment with inhibitors of protein or RNA synthesis or with inhibitors of caspases completely inhibited the arsenite effects. This result suggests that the arsenite activity requires new gene expression and activation of caspases. The arsenite was found to activate p38 and JNK3 but not JNK1 or JNK2 in the neurons. Blocking of p38 or JNK signaling pathways with inhibitors protected the neurons from arsenite toxicity. The authors suggest that the arsenite neurotoxicity is due to apoptosis caused by activation of p38 and JNK3 MAP kinases.

Daum et al. (2001) found that arsenite inhibits the Ras/ERK signaling pathway in smooth-muscle cells. The inhibition was attenuated by preincubation of the cells with N-acetylcysteine, suggesting that oxidative stress might be involved. The authors suggested that this mechanism could contribute to the known atherogenic actions of arsenical compounds. In contrast, Huang et al. (1999a) reported that arsenite at concentrations as low as 0.8 µM activates ERK phosphorylation in a mouse epidermal cell line, JB6 CL 41, used for tumor-promotion studies. At high doses (greater than 50 µM), arsenite also activated JNKs. Arsenite-induced ERK activation was blocked by introduction of dominant negative ERK-2 into cells. By also introducing a dominant negative form of JNK (JNK1) into these cells, the authors were able to demonstrate that activation of ERK, but not JNK, by arsenite is required to induce cell transformation. In another study, these same authors (Huang et al.1999b) found that 200 μM of arsenite induces both apoptosis and activates JNK in JB6 cells but did not induce p53-dependent transactivation. Furthermore, there was no difference in the induction of apoptosis by arsenite in p53$^{+/+}$ versus p53$^{-/-}$ cells, suggesting that induction of apoptosis at relatively high concentrations of arsenite proceeds via a p53-independent pathway.

Larochette et al. (1999) provided further evidence that arsenite acts on mitochondria to induce apoptosis, as mitochonidria were required for arsenite to induce nuclear apoptosis in a cell-free system. Arsenite caused the release of an apoptosis-inducing factor from the mitochondrial inner membrane space and also altered the permeability transition (PT) pore, suggesting that arsenite (30 μM) can induce apoptosis via a direct effect on the mitochondrial PT pore. In that study, arsenite induced apoptosis in a myelomonocytic leukemia cell line with an ED$_{50}$ (the effective dose in inducing apoptosis in 50% of the cells) of 20 μM.

Li and Broome (1999) found that 0.5-5 μM of As$_2$O$_3$ induces the expression of myeloid maturation and apoptotic markers, and markedly inhibits GTP-induced polymerization and microtubule formation. (However, only very high concentrations, 0.1-1 mM, were used.) The authors suggested that As^{III} binds to cysteine residues in tubulin, blocking the binding of GTP, thus disrupting the normal dynamic of microtubules during mitosis. They further suggested that that might be the mechanism by which As$_2$O$_3$ induced myeloid-cell maturation and arrest in metaphase, leading to apoptosis.

As^{III} is clearly cytotoxic to a variety of tumor cells in the 1-10 μM range; its toxicity is substantially influenced by the levels of intracellular glutathione (Yang et al. 1999). Using a variety of specific modulators of oxidative stress and cell signaling, Chen et al. (1998) found that arsenite-mediated apoptosis involved the formation of reactive oxygen species, activation of CPP32 (a protease) activity, PARP (a DNA repair enzyme) degradation, and release of cytochrome c from the mitochondria to the cytosol. From those and other data, the authors hypothesized that arsenite-induced apoptosis is triggered by the generation of H$_2$O$_2$ through activation of flavoprotein-dependent superoxide-producing enzymes such as NADH oxidase, which then might play a role as a mediator to induce apoptosis through activation of CPP32 protease, release of cytochrome c to cytosol, and PARP degradation.

Cell cultures were used to determine the effect of arsenicals on the proliferation and cytokine secretion of normal human epidermal keratinocytes (Vega et al. 2001). Trivalent arsenicals stimulated proliferation of the cells at concentrations of 0.001 to 0.01 μM, and concentrations greater than 0.5 μM inhibited proliferation. Pentavalent arsenicals did not stimulate proliferation. The trivalent arsenicals also stimulated secretion of the growth-promoting cytokines, granulocyte macrophage colony-stimulating factor, and tumor necrosis factor. The study indicated possible mechanisms by which arsenicals can cause damage to human skin or contribute to neoplasia. Those observations are consistent with earlier data showing a stimulatory effect on proliferation of keratinocytes at low doses (Germolec et al. 1997, 1998), as discussed in the 1999 NRC report.

Sodium arsenite also reduced the proliferation of PHA-stimulated T lymphocytes from human peripheral blood due to inhibition of secretion of IL-2 (Vega et al. 1999).

A preliminary report presented at a meeting by Chiang et al. (2001) found that continuous exposure of immortalized human keratinocytes (HaCaT) to noncytotoxic doses of arsenite (0.5 and 1.0 μM) for 5 months resulted in increased cell-culture growth rate, cell density, and colony-forming efficiency. The arsenite-exposed HaCaT cells formed tumors when injected into nude mice, in contrast to the absence of tumor formation following injection of unexposed cells into nude mice.

Induction of Oxidative Stress

Prior to the 1999 NRC report, a great deal of research was conducted on the genotoxic and chromosomal effects of arsenic. Recently, several studies have provided additional evidence that arsenic can induce DNA damage indirectly by promoting the formation of oxygen free radicals.

Barchowsky et al. (1999a) found that 5-200 μM of arsenite caused cell proliferation and increased superoxide and H_2O_2 accumulation, increased activity of the cytoplasmic tyrosine kinase cSRC, increased H_2O_2-dependent tyrosine phosphorylation, and increased NF-κB-dependent transcription. At higher concentrations that caused cell death, As[III] activated MAP kinases and p38. Another study by this group (Barchowsky et al. 1999b) demonstrated via electron spin resonance that these same concentrations of As[III] rapidly increased oxygen consumption and superoxide formation, which was abolished by superoxide dismutase (SOD) but not nitric oxide radicals, confirming that superoxide is a predominant form of arsenic-induced reactive oxygen species. The authors suggested from this finding that arsenite-induced oxidant accumulation, in the form of superoxide and H_2O_2, activates tyrosine phosphorylation and might represent a MAP-kinase-independent pathway for proliferation of vascular cells. Lynn et al. (1998) found that arsenite-induced poly(ADP)-ribosylation, NAD depletion, DNA-strand breaks, and micro-nuclei formation in CHO-Ki cells. However, in contrast to the findings of Barchowsky et al. (1999a), Lynn et al. (1998) noted that the DNA damage could be suppressed by pretreatment with inhibitors of nitric oxide synthetase and also found increased nitrite levels in cells following As^{III} treatment. From those results, they concluded that the induction of nitric oxide might be important to the etiology of arsenic-induced vascular disorders in humans.

The mechanisms by which arsenic causes oxidant stress were further elucidated in several in vitro studies. The effects of four species of arsenic (arsenite, arsenate, MMA, and DMA) in mobilizing iron from horse-spleen ferritin were investigated (Ahmad et al. 2000). The results indicate that exogenous methylated arsenic species administered alone or synergistically in the presence of endogenous ascorbic acid can cause the release of iron from ferritin and an iron-dependent formation of reactive oxygen species.

In rats, Kitchin et al. (1999) found a linear relationship between the tissue concentration of arsenite and the induction of heme oxygenase (HO), a metabolic enzyme known to be induced by stress. Doses resulting in 30 μmol of arsenite per kilogram of liver induced HO, while a dose resulting in 10 μmol/kg of liver did not induce HO. The tissue concentration of arsenite needed to induce kidney HO was 100 μmol/kg of kidney; 30 μmol/kg did not induce HO.

The ability of different forms of arsenic to inhibit thioredoxin reductase, a nicotinamide adenine dinucleotide phosphate (NADPH)-dependent flavoenzyme that catalyzes the reduction of many disulfide-containing substrates and plays an important role in the cellular response to oxidative stress, was investigated by Lin et al. (2001). Of the arsenicals tested (As^{III}, MMA^{III}, and DMA^{II}), the monomethyl form of trivalent arsenic was found to be the most potent inhibitor of thioredoxin reductase, having an IC_{50} of 3 μM.

S.X. Liu et al. (2001) used a fluorescent probe sensitive to oxygen radicals and a confocal scanning microscope to show that arsenite is capable of rapidly increasing intracellular levels of oxygen radicals. The arsenic-induced increase in oxygen radicals could be reduced by the free radical scavenger, dimethyl sulfoxide (DMSO). Using electron spin resonance spectroscopy and selective spin-trap agents, the authors also showed that arsenite can increase the levels of superoxide-derived hydroxyl radical production in a Chinese hamster ovary cell line that contains a single copy of human chromosome 11. Finally, they demonstrated that arsenite could induce mutations in the CD59 gene on chromosome 11 in these cells. The mean lethal dose of arsenite to these cells was 1.7 μg/mL (23 μM) but was 0.1 μg/mL (2.3 μM) following glutathione depletion with buthionine sulfoximine, indicating that cell death was mediated via oxidative stress. These results suggest that arsenite can produce intracellular oxygen free radicals capable of damaging DNA.

Matsui et al. (1999) examined the DNA in human skin biopsies obtained from cases of Bowen's disease (a type of skin cancer), invasive squamous-cell carcinoma (SCC), and arsenical keratoses for the frequency of 8-hydroxy-2-deoxyguanosine, a marker of oxidative damage. Cases were obtained from unspecified areas of Taiwan, Japan, and Thailand where the authors indicated "chronic arsenicism was endemic." Control biopsies consisted of 10 Bowen's disease lesions and one SCC obtained from 11 Japanese subjects without known arsenic exposure. The authors found that the frequency of oxidative damage to DNA was significantly higher (22 of 28; 78%) in the arsenic-related lesions when compared with those from subjects not exposed to arsenic (1 of 11; 9%). For a subset of these samples, arsenic exposure was determined by neutron activation analysis of deparaffined skin-tumor samples. Arsenic was readily identified in four of five tumors classified as arsenic-related, but not in any of three cases of non-arsenic-related tumors. Those data provide further evidence that arsenic is capable of causing oxidative damage to DNA and suggest that this mechanism could contribute to the development of arsenic-related skin tumors in humans exposed to arsenic through drinking water.

Lynn et al. (2000) found that treatment of human vascular smooth-muscle cells in culture with concentrations of arsenite greater than 1 μmol/L increased superoxide concentrations and induced oxidative damage to DNA. The authors also provided evidence suggesting that low concentrations of arsenite can activate NADH oxidase, which could contribute to the formation of super-oxide.

Studies by Snow and colleagues (Snow et al. 1999, in press; Chouchane and Snow 2001) in human keratinocytes indicated the different effects of As^{III} at different concentrations. Ligase function, a process involved in DNA repair, was increased in SV-40 transformed cells (AG06) after a 3-hr exposure with approximately 10 μM of arsenite and was decreased in a dose-dependent manner after 24-hr exposures with arsenite concentrations up to 100 μM (Snow et al. 1999). However, the purified ligases were only inhibited at very high (millimolar) concentrations of arsenite. The authors suggest that the inhibition of DNA repair by arsenite is by an indirect mechanism, such as transcriptional or post-translational regulation of gene products.

At micromolar concentrations of arsenite (maximal effect at 3 μM), there was an increase in cellular GSH due to an increase in the activity of the GSH-synthesizing enzyme gamma-glutamyl cysteinyl synthase. GSH and N-acetyl-cysteine were found to protect against arsenite toxicity. However, the various enzymes that use GSH as a substrate (e.g., glutathione transferases, glutathione reductase, and glutathione peroxidase) were not inhibited by relevant concentrations of arsenite (Snow et al. 1999; Chouchane and Snow 2001). By contrast, the enzyme pyruvate dehydrogenase, an enzyme involved in energy production, was inhibited with an IC_{50} of 5.6 μM of arsenite, and this inhibition is likely to be involved in the cytotoxicity of arsenic in the cells.

In addition to the effects on ligase function and cellular redox enzymes, arsenite was found to interact with DNA alkylating agents. The activity of low concentrations of arsenite was opposite to that of high concentrations. Cells pre-exposed for 24 hr with 0.2 μM of arsenite before exposure to less than 4 μM of N-methyl-N-nitro-N-nitrosoguanidine (MNNG) showed an enhanced uptake of neutral red dye compared with that of either exposure alone. (Dye uptake increased rather than decreased, a sign of either increased lysosomal activity or cell proliferation.) In contrast, pre-exposure with 4 or 20 μM of arsenite resulted in decreased uptake of neutral red dye compared with that seen with 20 to 35 μM of MNNG. (Neutral red dye decreased, a sign of decreased cell viability or cytotoxicity.)

Alterations in Gene Expression

In the 1999 NRC report, alterations in gene expression via hypo- or hypermethylation of CpG sites in DNA were reviewed. Recently, Zhong and Mass (2001) utilized a sensitive PCR technique to determine whether inorganic As^{III} can alter DNA methylation at relatively low concentrations in immortalized human kidney tumor cells in culture. The kidney cell line was considerably more susceptible to cytotoxic effects of arsenic (IC_{50} = 20 nM) than the human lung cell line, A549 cells (IC_{50} = 400 nM). Those relatively low concentrations of arsenic in cell culture produced six differentially hypermethylated regions and two differentially hypomethylated regions of genomic DNA, lending further support to the hypothesis that some forms of arsenic can alter gene regulation through hyper- or hypomethylation of DNA. Although this group of investigators had shown that hypermethylation of CpG sites can inactivate the transcriptional activity of human p53 tumor suppressor gene (Schroeder and Mass 1997) and that arsenic can alter methylation patterns in the promoter region of the p53 gene (Mass and Wang 1997), Zhong and Mass (2001) failed to find arsenic-induced alterations in methylation of the CpG-sensitive promoter of the von Hippel-Lindau tumor suppressor gene in their recent study.

Recent studies have examined p53 gene expression and mutation in tumors obtained from subjects with a history of arsenic ingestion. Kuo et al. (1997) compared p53 overexpression in 26 cases of Bowen's disease lesions from residents in the area of southwestern Taiwan where blackfoot-disease is endemic presumably exposed to arsenic and in 22 cases of Bowen's disease lesions from unexposed control subjects. Immunohistochemical staining for p53 occurred at a positive rate in biopsies from 42.3% (11 of 26) arsenic-exposed patients compared with 9.1% (2 of 22) from nonexposed patients (p = 0.01). In another study from the region where blackfoot-disease is endemic, Chang et al. (1998) reported positive immunohistochemical staining for p53 in 100% of 30 Bowen's disease lesions, 12 basal-cell carcinomas (BCCs), and 10 SCCs. Normal skin adjacent to the lesions also consistently overexpressed p53. In a study by Hsu et al. (1999) of skin cancer in subjects from the region where blackfoot-disease is endemic, immunohistochemical staining found p53 overexpression in 43.5% (10 of 23) of lesions of Bowen's disease, in 14% (1 of 7) of the BCCs, and in 44% (4 of 9) of the SCCs. Polymerase chain-reaction single-strand conformation polymorphism (PCR-SSCP) analysis detected p53 gene mutations in 39% (9 of 23) of the cases with Bowen's disease, 28.6% (2 of 7) of cases with BCC, and 55.6% of cases with SCC. The two techniques did not have complete concordance: PCR-SSCP analysis did not detect mutations in 4 of 15 cases with positive immunohistochemical staining and did detect mutations in 5 of 15 cases that stained negative. In a recent study of BCC from Queensland, Australia, Boonchai et al. (2000) found a lower percentage of p53 expression in biopsies obtained from patients with a history of prior chronic ingestion of an arsenic-containing elixir compared with those with no known arsenic exposure and presumed UV-induced basal-cell carcinoma (52.9% of 121 biopsies).

Hamadeh et al. (1999) found that treatment of human keratinocytes to low concentrations of As^{III} (1-1,000 nM) produced a dose-dependent increase in mdm2 protein (involved in p53 regulation) and a corresponding decrease in p53 protein. Effects were seen at all time points from 2 to 14 days of incubation, but peaked at about 4-6 days. The authors proposed that disruption of the *p53-mdm2* loop that regulates cell-cycle arrest might be an important event in the development of arsenic-related skin carcinogenesis.

Several other recent studies provide additional support for the hypothesis that arsenic can modulate gene expression. Kaltreider et al. (1999) found that relatively low (1 μM), noncytotoxic doses of arsenite alter nuclear-binding levels of several transcription factors, including AP-1, NF-kB, SP1, and Yb-1, to their cis-acting elements in a human breast cancer cell line and in rat hepatoma cells. A subsequent study demonstrated that 3.3 μM of arsenite altered the function of the nuclear glucocorticoid receptor, apparently by interacting directly with the glucocorticoid receptor complex and interfering with GR-mediated transcriptional activation of genes in dexamethasone-treated, hormone-responsive H4IIE rat hepatoma cells (Kaltreider et al. 2001). This type of toxicity might influence not only the induction of neoplasia but could also influence other corticosteroid-dependent activities.

Parrish et al. (1999) investigated the impact of submicromolar concentrations of arsenite and arsenate on transcription-factor DNA binding and gene expression in precision-cut rabbit renal cortical slices. Arsenite and arsenate appeared equipotent in several effects, including increasing AP-1 DNA-binding activity and increasing c-*fos*, c-*jun*, and c-*myc* gene expression. Peak effects for both arsenic species occurred in the range of 0.1 to 1.0 μM. Arsenite, but not arsenate, increased expression of heat-shock protein HSP 32 at submicromolar concentrations. This study extends earlier observations of the impact of low-dose arsenite on gene expression—for example, the finding by Germolec et al. (1997) that 1 μM of arsenite increased expression of the c-*myc* protooncogene in normal human keratinocytes in vitro.

Tully et al. (2000) used a battery of recombinant HepG2 cells (a human liver tumor-derived cell line) to evaluate the effects of metals, including As^V, on transcriptional activation of any of 13 signal transduction pathways. At high doses (100 μM and greater), they found that As^V induced transcriptional activation of several genes, including glutathione S-transferase A1, metallothionein IIA, NF-κB, FOS, and HSP70. However, because effects were not generally seen at the lowest concentration tested (50 μM), the relevance of these results to risk assessment of low concentrations of arsenic in drinking water is limited.

Lu et al. (2001) used cDNA microarrays to evaluate whether gene expression in six needle biopsies of human liver tissue from individuals exposed to arsenic in their environment from the Guizhou region of China was different from normal human liver. Chronic exposure to arsenic apparently occurs from the use of coal containing very high concentrations of arsenic (up to 900 ppm) for the purposes of drying food. All cases were selected from subjects with a history of arsenic-related skin lesions (e.g., keratoses and hyperpigmentation) and gastrointestinal disturbances. Furthermore, all of the Chinese patients had histopathological evidence of chronic liver damage that was presumed to be related to their arsenic exposure. Patients did not have a history of hepatitis, and serology for hepatitis B and C viruses was negative. When compared with normal liver obtained from six surgical resections or transplantations, approximately 60 genes of about 600 examined appeared to be differentially expressed. Of particular interest were changes in expression of genes involved in cell-cycle regulation, cell proliferation, apoptosis, and DNA-damage repair. Although there are many uncertainties associated with the design and interpretation of this human study, the authors noted that the observed changes in gene expression were consistent with the result of array analysis of chronically arsenic-exposed mouse livers and chronically arsenic-transformed rat liver cells seen in other studies (Chen et al. 2001; Liu, J. et al. 2000).

In another global gene expression study, Liu et al. (2001a) found that mice exposed to high doses of inorganic arsenicals As^{III} at 100 μmol/kg or As^V at 300 μmol/kg exhibited significant changes in expression of genes related to stress (e.g., HSP60, heme oxygenase I), DNA damage (e.g., GADD45 and the DNA excision repair protein ERCC1), and xenobiotic biotransformation (e.g., various cytochrome P-450s). They also found that arsenic exposure activated the c-Jun/AP-1 transcription complex, which could explain some of the widespread changes in gene expression.

Collectively, these studies provide further evidence that various forms of arsenic can alter gene expression and that such changes could contribute substantially to the toxic and carcinogenic actions of arsenic in human populations.

Resistance and Tolerance to Arsenic Cytotoxicity

Romach et al. (2000) investigated the mechanism underlying the development of tolerance to arsenic in a rat liver epithelial cell line (TRL 1215). That cell line over-expresses metallothionein (MT), a protein that binds metals and is often associated with the development of tolerance to metals. The cells were chronically (18-20 weeks) exposed to 500 nM of arsenite. Control cells (not previously exposed to arsenite) and the chronically exposed cells were subsequently exposed to arsenite. The LC_{50} in the chronically exposed cells was 140 μM compared with 26 μM in the control cells. The LC_{50}s for As^V and DMA, but not MMA, were also increased. The control cells, however, were extremely tolerant to MMA to begin with ($LC_{50} > 60,000$ μM). Culturing the chronically exposed cells in arsenite-free media did not alter their tolerance, suggesting an irreversible phenotypic change. The development of tolerance was not linked to GSH levels nor metallothionein levels in the cells.

The tolerant cells accumulated less arsenite than the control cells, had an increased methylation of arsenic to DMA, and eliminated the arsenite more readily. The increased methylation could not fully account for the increased tolerance, and the mechanism underlying the development of tolerance is still unclear. It should be noted that the observed tolerance is to the cytotoxic effects of arsenic, but that does not necessarily imply tolerance to the carcinogenic effects of arsenic.

Rossman and Wang (1999) identified two cDNAs, asr1 and asr2, which confer As[III] resistance to As[III]-sensitive cells lines. In a second study, Rossman and Wang (1999) used arsenite-sensitive and arsenite-resistant strains of Chinese hamster V79 cells and found that overexpression of the tumor suppressor gene *fau* contributed to arsenic resistance (Rossman and Wang 1999). They hypothesized that because *fau* contains a ubiquitin-like region, arsenic might interfere with the ubiquitin system that targets certain proteins for rapid destruction. Because p53 and several other proteins involved in DNA repair and cell-cycle regulation also require ubiquitin signal sequences for proper function, Rossman and Wang (1999) suggested that the carcinogenic effects of arsenite might, in part, be mediated indirectly by effects on cellular control of DNA repair processes, secondary to interference with the ubiquitin system.

Liu et al. (2001b), who earlier found that long-term exposure to low concentrations of arsenite-induced malignant transformation in a rat-liver epithelial cell line, reported that these chronically exposed cells developed a tolerance to later acute exposures. Microarray analysis of the cells showed increased expression of the genes encoding for glutathione *S*-transferase, multidrug resistance-associated protein genes, and multidrug resistance genes. The tolerant cells were found to be resistant to several anticancer drugs.

Implications of Mode-of-Action Studies of Arsenic in Drinking Water to Human Carcinogenesis

As described in this chapter, numerous recent studies have identified potential modes of action by which arsenic could increase cancer risk in human populations. Several recent in vivo animal studies have been completed, including one preliminary study that potentially demonstrated a significant carcinogenic response in mice to arsenic at 500 ppb (500 µg/L) in drinking water. However, the lack of dose-response information and other uncertainties in extrapolating from rodents to humans makes rodent bioassays unsuitable for direct application to human risk assessment for arsenic, particularly in view of the extensive human epidemiological data that are available. Since the 1999 NRC report, several studies have reported the presence of trivalent methylated arsenic metabolites in tissues or urine. Because these forms are reactive and highly toxic, they have been considered in the evaluation of possible modes of action of arsenic carcinogenicity.

Recent studies have used in vitro experiments to evaluate several putative modes of action for arsenic-induced cancer, including (1) induction of mutations and chromosomal abberations, (2) secondary induction of DNA damage via generation of oxygen free radicals (oxidative stress) and altered fidelity of DNA repair, (3) alterations in signal transduction, cell-cycle regulation, and apoptosis, and (4) alterations in gene expression. The newly reported studies on mode of action of arsenic provide insight into the importance of exposure level, exposure duration, and time frame of observation in the assessment of cellular responses to arsenic. Within the same cell system, different levels of exposure appeared to induce different modes of action (Table 3-1).

To assess the potential relevance and implications of these in vitro studies to the risk assessment of arsenic in drinking water, it would be useful to determine whether the studies used experimental concentrations of arsenic that are plausible in human target tissues following exposure via drinking water. Human urinary bladder epithelium is one of the primary target tissues for arsenic carcinogenesis. Because this tissue is essentially bathed in urine that contains arsenic following ingestion of arsenic in the diet and drinking water, levels of biologically active forms of arsenic that appear in the urine following exposure to low-to-moderate doses in humans are particularly relevant to in vitro studies. Numerous studies have speciated arsenic in urine following ingestion via drinking water. The mechanistic studies reviewed herein and those reviewed previously in the 1999 NRC report suggest that trivalent arsenic species (primarily As^{III}, MMA^{III}, and, possibly, DMA^{III}) are the forms of arsenic of greatest toxicological concern. Considering a concentration of arsenic in drinking water of 50 μg/L (i.e., 50 ppb) for illustrative purposes, the corresponding concentration of As^{III} in urine can be estimated from previous studies. In general, arsenic at 50 μg/L of drinking water would be expected to yield a total urinary arsenic concentration in the range of 30-100 μg/L (NRC 1999). Most studies have found that approximately 10-30% of total urinary arsenic is inorganic (As^{III} and As^{V}) (NRC 1999), yielding urinary concentrations of inorganic arsenic in the range of 3-30 μg/L. Approximately 50-75% of the inorganic arsenic concentration is As^{III} (Smith et al. 1977; Yamamura and Yamauchi 1980; Farmer and Johnson 1990; Hakala and Pyy 1995; Crecelius and Yager 1997), yielding a concentration of As^{III} at approximately 2-24 μg/L, equivalent to 0.04-0.3 μM. Because MMA^{III} and DMA^{III}

might also be toxicologically relevant and have been found as metabolites in urine, some adjustment upward of this concentration would be appropriate for comparative purposes. It should be noted that background exposure to arsenic commonly results in urinary concentrations of inorganic arsenic of around 1-2 µg/L (Foa et al. 1984; Kalman et al. 1990; Lin and Huang 1995), yielding a baseline concentration of approximately 0.02-0.04 µM of As^{III}. Therefore, in assessing the relevance of in vitro studies to arsenic-induced cancer, the following guidelines seem appropriate:

1. Arsenite concentrations of 0.01-0.5 µM approximate the concentrations that might exist in human urine following ingestion of drinking water containing arsenic concentrations of 3-50 ppb, a range that is currently the focus of low- dose risk assessment.

2. Arsenite concentrations of 0.5-10 µM have been encountered in the urine of human populations chronically exposed to higher concentrations (more than 100 µg/L) of arsenic in drinking water and might be relevant to understanding pathological effects observed in human epidemiological studies.

3. Arsenite concentrations in excess of 10 µM generally exceed concentrations that can occur in the urine of individuals chronically exposed to arsenic in drinking water and have less direct relevance to understanding the modes of action responsible for human cancer induced by this route of exposure.

As discussed in this chapter, several recent studies have identified biological effects of arsenic at in vitro concentrations of less than 1 µM; although for many others, high concentrations (greater than 10 µM) were required to produce biological effects. The following biochemical effects have been seen in in vitro studies at arsenic concentrations of 1 µM or less, with a few showing effects at concentrations less than 0.1 µM (see Table 3-1):

1. Induction of oxidative damage to DNA.
2. Altered DNA methylation and gene expression.[2]
3. Changes in intracellular levels of mdm2 protein and p53 protein.[2]
4. Inhibition of thioredoxin reductase (MMA^{III}, but not As^{III}).
5. Inhibition of pyruvate dehydrogenase.
6. Altered colony-forming efficiency.[2]

---

[2]These biological effects were observed at concentrations of less than 0.1 µM of arsenic.

7. Formation of protein-DNA cross-links.[2]
8. Induction of apoptosis.[2]
9. Altered regulation of DNA repair genes, thioredoxin, glutathione reductase, and other stress-response pathways.
10. Stimulation (0.001-0.01 μM) and inhibition (greater than 0.5 μM) of normal human keratinocyte cell proliferation.[2]
11. Altered function of the glucocorticoid receptor.

Some of the effects described above would act to promote carcinogenic activity (e.g., stimulation of cell proliferation), while others would act to prevent cancer formation (e.g., apoptosis). The results of the mode-of-action studies do not provide a clear picture of the shape of the dose-response curve at low doses.

At concentrations between 1 and 10 μM, a variety of additional biochemical effects have been observed. For example, induction of apoptosis has been seen in numerous studies and might be an important, if not essential, mechanism for the chemotherapeutic effects of arsenic toward acute promyelocytic leukemia and perhaps certain other types of cancer cell lines studied in vitro.

Cytogenetic studies of malignant and nonmalignant tissues obtained from humans exposed to arsenic by ingestion have found evidence of genotoxic effects, including DNA fragmentation, chromosomal aberrations and micronuclei, alterations in gene expression, and specific gene mutations (notably in p53). The relatively few studies conducted to date have examined populations exposed to concentrations of ingested arsenic associated with urinary inorganic arsenic concentrations in the range of 0.1-5 μM.

Thus, evidence is accumulating that relatively low concentrations of arsenic, potentially achievable through consumption of drinking water containing 10-50 μg/L of arsenic, can alter biochemical pathways relevant to carcinogenesis (gene expression, cell-cycle regulation, signal transduction, apoptosis, oxidative stress, and others). Before drawing conclusions on the relevance of these effects, however, one must be aware of the many caveats present in extrapolating from in vitro studies. For example, effects seen in transformed epithelial cell lines might or might not be relevant to normal epithelial-cell responses because of the many biochemical differences that are present in transformed cells. This is especially true for tumor cell lines derived from different primary tumors. Effects seen in one cell type might or might not be relevant to other cell types because of differences among cell types in uptake and metabolism of arsenic, differences in gene expression, and differences in redox pathways. For example, effects seen in human keratinocytes might or might not be relevant to effects in human bladder or liver epithelial cells and vice versa. It is important to note that such differences might make the in vitro model either less or more sensitive to arsenic, relative to in vivo exposures. There is no way of knowing a priori whether an effect of arsenic at a given concentration seen in one cell type in vitro is directly relevant to target-tissue effects in vivo following arsenic exposure via drinking water. In addition, the uptake and efflux of arsenic and its metabolites might be different in in vitro experiments. Of course, all of the uncertainties associated with cross-species comparisons are also relevant to in vitro studies using cell lines or tissues from different species.

All of the above modes of action (i.e., chromosomal alterations, alterations in DNA caused by oxidative stress or by altered methylation, and inhibition of DNA-repair enzymes) could enhance the carcinogenic actions of other direct tumorigens by loss of efficient repair processes or loss of control elements in the genetic material. Thus, the mode-of-action studies suggest that the arsenic might be acting as a cocarcinogen, a promoter, or a progressor.

The aforementioned biological effects that might occur following exposure to arsenic are likely to exhibit a typical sigmoid dose-response relationship, characterized by a sublinear, linear, or supralinear shape, depending on where the dose range under consideration falls along the dose-response curve. An important and controversial issue is whether the existing mode-of-action data provide suitable evidence to demonstrate that arsenic acts as a threshold-type carcinogen in humans exposed to arsenic in drinking water. A sublinear response might imply the existence of a threshold below which there is no biologically relevant response. It should be recognized that the actual shape of the dose-response curve seen in vivo following chronic exposure to arsenic in drinking water will be a composite of many individual dose-response curves for different specific biochemical end points. It is likely that multiple different cellular perturbations are necessary before a pathological end point, such as tumor development, occurs. Thus, inferring the shape of the in vivo dose-response curve from mechanistic studies cannot be done with any confidence.

Despite the publication of important new findings since 1999, it is the subcommittee's consensus that the existing research database cannot establish with confidence the precise modes of action involved in arsenic-induced cancer in humans. The shape of the dose-response relationship observed for a candidate biochemical effect in a specific in vitro model should not imply that the overall dose-response relationship for the appearance of cancer in exposed human populations is likely to have the same or similar shape. At the present time, the multiplicity of potential modes of action and the apparent heterogeneity of susceptibility and responsiveness that characterizes human populations does not allow identification of a specific range of arsenic concentrations in drinking water for which the occurrence of cancer would be expected to exhibit a nonlinear as opposed to a linear dose-response relationship.

SUMMARY AND CONCLUSIONS

• Inorganic arsenic is methylated via alternating reduction of pentavalent arsenic to trivalent arsenic and addition of a methyl group from *S*-adenosylmethionine. The main metabolic products MMA^V and DMA^V are readily excreted in urine. There is increasing evidence for the presence of reactive amounts of trivalent methylated arsenic metabolites, especially MMA^{III}, in tissues and urine following exposure to inorganic arsenic. Therefore, the methylation of inorganic arsenic is not entirely a detoxification process; the formation and distribution of the reduced metabolites might be associated with increased toxicity. Those metabolites can be released and excreted in urine, especially when complexed with dithiols like DMPS. The exact mechanisms of arsenic methylation are not known. In particular, the arsenic methyltransferases and reductases are only partially characterized. Arsenic methylation seems to vary considerably between tissues. DMA is the main metabolite excreted from the cells. There are major differences in the methylation of inorganic arsenic among mammalian species; therefore, the results of studies on animals or animal cells are difficult to extrapolate to humans.

• Some recent experiments in animals have demonstrated an increase in cancer incidence following exposure to inorganic arsenic or dimethylarsenic acid either alone or in the presence of an initiator. However, in view of the extensive human epidemiological data that are available, those studies, although qualitatively relevant, are not appropriate for use in a quantitative human-health risk assessment for arsenic.

• Experiments in animals and in vitro have demonstrated that arsenic has many biochemical and cytotoxic effects at low doses and concentrations that are potentially attainable in human tissues following ingestion of arsenic in drinking water. Those effects include induction of oxidative damage to DNA; altered DNA methylation and gene expression; changes in intracellular levels of mdm2 protein and p53 protein; inhibition of thioredoxin reductase (MMA^{III} but not As^{III}); inhibition of pyruvate dehydrogenase; altered colony-forming efficiency; induction of protein-DNA cross-links; induction of apoptosis; altered regulation of DNA- repair genes, thioredoxin, glutathione reductase, and other stress-response pathways; stimulation or inhibition of normal human keratinocyte cell proliferation, depending on the concentration; and altered function of the glucocorticoid receptor.

- Despite the extensive research investigating the modes of action of arsenic, the experimental evidence does not allow confidence in distinguishing between various shapes (sublinear, linear, or supralinear) of the dose-response curve for tumorigenesis at low doses. Therefore, the choice of model to extrapolate human epidemiological data from the observed range (100-2,000 μg/L) to the range of regulatory interest (3-50 μg/L) cannot be made solely on a mechanistic basis.

- Although a threshold for carcinogenic activity for arsenic might conceivably exist in a given individual, interindividual variation in response and the variability in background exposure to arsenic via the diet and other sources within the human population complicates the determination of a no-effect level in a diverse human population, if such a level were to exist.

RECOMMENDATIONS

- Research should be conducted to determine in vivo target-tissue concentrations of the various arsenic metabolites following ingestion of arsenic in drinking water. Of particular interest are the trivalent arsenic metabolites.

- Additional information is needed on arsenic reduction and methylation reactions and on reactions of the various metabolites with critical tissues and proteins. Studies on the metabolism of arsenic should preferably be conducted in humans or in human cells.

- Future research into the various potential biochemical and molecular modes of action of arsenic should focus on identifying the dose-response relationship and time course for effects of arsenic at the submicromolar level in human-derived cells or tissues.

- The inclusion of potential biomarkers of cytogenetic effect in epidemiological studies of arsenic-induced cancer help to elucidate the mechanisms of arsenic carcinogenesis and ultimately increase the power of investigations conducted in populations with low-dose exposure.

- Research should be conducted to determine the extent to which arsenic acts as a cocarcinogen with known carcinogens.

REFERENCES

Ahmad, S., W.L. Anderson, and K.T. Kitchin. 1999. Dimethylarsinic acid effects on DNA damage and oxidative stress related biochemical parameters in B6C3F1 mice. Cancer Lett. 139(2):129-135.

Ahmad, S., K.T. Kitchin, and W.R. Cullen. 2000. Arsenic species that cause release of iron from ferritin and generation of activated oxygen. Arch. Biochem. Biophys. 382(2):195-202.

Alemany, M., and J. Levin. 2000. The effects of arsenic trioxide (As2O3) on human megakaryocytic leukemia cell lines with a comparison of its effects on other cell lineages. Leuk. Lymphoma 38(1-2):153-163.

Anundi, I., J. Högberg, and M. Vahter. 1982. GSH release in bile as influenced by arsenite. FEBS Lett. 145(2):285-288.

Aposhian, H.V., E.S. Gurzau, X.C. Le, A. Gurzau, S.M. Healy, X. Lu, M. Ma, L. Yip, R.A. Zakharyan, R.M. Maiorino, R.C. Dart, M.G. Tircus, D. Gonzalez-Ramirez, D.L. Morgan, D. Avram, and M.M. Aposhian. 2000b. Occurrence of mono-methylarsonous acid in urine of humans exposed to inorganic arsenic. Chem. Res. Toxicol. 13(8):693-697.

Aposhian, H.V., B. Zheng, M.M. Aposhian, X.C. Le, M.E. Cebrian, W. Cullen, R.A. Zakharyan, M. Ma, R.C. Dart, Z. Cheng, P. Andrewes, L. Yip, G.F. O'Malley, R.M. Maiorino, W. Van Voorhies, S.M. Healy, and A. Titcomb. 2000a. DMPS—Arsenic challenge test. II. Modulation of arsenic species, including monomethylarsonous acid (MMA^{III}), excreted in human urine. Toxicol. Appl. Pharmacol. 165(1):74-83.

ATSDR (Agency for Toxic Substances and Disease Registry). 2000. Toxicological Profile for Arsenic. U.S. Department of Health and Human Services, Agency for Toxic Substances and Disease Registry, Atlanta, GA. [Online]. Available: http://www.atsdr. cdc.gov/toxprofiles/tp2.html [August 15, 2001].

Barchowsky, A., R.R. Roussel, L.R. Klei, P.E. James, N. Ganju, K.R. Smith, and E.J. Dudek. 1999a. Low levels of arsenic trioxide stimulate proliferative signals in primary vascular cells without activating stress effector pathways. Toxicol. Appl. Pharmacol. 159(1):65-75.

Barchowsky, A., L.R. Klei, E.J. Dudek, H.M. Swartz, and P.E. James. 1999b. Stimulation of reactive oxygen, but not reactive nitrogen species, in vascular endothelial cells exposed to low levels of arsenite. Free Radic. Biol. Med. 27(11-12):1405-1412.

Biggs, M.L., D.A. Kalman, L.E. Moore, C. Hopenhayn-Rich, M.T. Smith, and A.H. Smith. 1997. Relationship of urinary arsenic to intake estimates and a biomarker of effect, bladder cell micronuclei. Mutat. Res. 386(3):185-195.

Boonchai, W., M. Walsh, M. Cummings, and G. Chenevix-Trench. 2000. Expression of p53 in arsenic-related and sporadic basal cell carcinoma. Arch. Dermatol. 136(2):195-198.

Brown, J.L., K.T. Kitchin, and M. George. 1997. Dimethylarsinic acid treatment alters six different rat biochemical parameter arsenic carcinogenesis. Teratog. Carcinog. Mutagen. 17(2):71-84.

Chang, C.H., R.K. Tsai, G.S. Chen, H.S. Yu, and C.Y. Chai. 1998. Expression of bcl-2, p53 and Ki-67 in arsenical skin cancers. J. Cutan. Pathol. 25(9):457-462.

Chen, H., J. Liu, B.A. Merrick, and M.P. Waalkes. 2001. Genetic events associated with arsenic-induced malignant transformation: applications of cDNA microarray technology. Mol. Carcinog. 30(2):79-87.

Chen, N.-Y., W.Y. Ma, C. Huang, M. Ding, and Z. Dong. 2000a. Activation of PKC is required for arsenite-induced signal transduction. J. Environ. Pathol. Toxicol. Oncol. 19(3):297-306.

Chen, H., S. Qin, and Q. Pan. 2000b. Antitumor effect of arsenic trioxide on mice experimental liver cancer [in Chinese]. Zhonghua Gan Zang Bing Za Zhi 8(1):27-29.

Chen, T., Y. Na, and S. Fukushima. 1999. Loss of heterozgosity in (LewisxF344)F1 rat urinary bladder tumors induced with N-butyl-N-(4-hydroxybutyl) nitrosamine followed by dimethylarsinic acid or Sodium L-ascorbate. Jap. J. Cancer Res. 90(8):818-823.

Chen, Y.C., S.Y. Lin-Shiau, and J.K. Lin. 1998. Involvement of reactive oxygen species and caspase 3 activation in arsenite-induced apoptosis. J. Cell Physiol. 177(2):324-333.

Chiang, M.C., L.H. Yih, R.N. Huang, K. Peck, and T.C. Lee. 2001. Tumor formation of immortalized HaCaT cells in nude mice by long-term exposure to sodium arsenite at non-toxic doses. Toxicology 164(1-3):95(P2A21)

Chouchane, S., and E.T. Snow. 2001. In vitro effect of arsenical compounds on glutathione-related enzymes. Chem. Res. Toxicol. 14(5):517-522.

Cohen, S.M., S. Yamamoto, M. Cano, and L.L. Arnold. 2001. Urothelial cytotoxicity and regeneration induced by dimethylarsinic acid in rats. Toxicol. Sci. 59(1):68-74.

Crecelius, E., and J. Yager. 1997. Intercomparison of analytical methods for arsenic speciation in human urine. Environ. Health Perspect. 105(6):650-653.

Cullen, W.R., and K.J. Reimer. 1989. Arsenic speciation in the environment. Chem. Rev. 89(4):713-764.

Daum, G., J. Pham, and J. Deou. 2001. Arsenite inhibits Ras-dependent activation of ERK but activates ERK in the presence of oncogenic Ras in baboon vascular smooth muscle cells. Mol. Cell Biochem. 217(1-2):131-136.

De Kimpe, J., R. Cornelis, and R. Vanholder. 1999a. In vitro methylation of arsenite by rabbit liver cytosol: effect of metal ions, metal chelating agents, methyltransferase inhibitors and uremic toxins. Drug. Chem. Toxicol. 22(4):613-628.

De Kimpe, J., R. Cornelis, L. Mees, R. Vanholder, and G. Verhoeven. 1999b. 74As-arsenate metabolism in Flemish Giant rabbits with renal insufficiency. J. Trace Elem. Med. Biol. 13(1-2):7-14.

Deaglio, S., D. Canella, G. Baj, A. Arnulfo, S. Waxman, and F. Malavasi. 2001. Evidence of an immunologic mechanism behind the therapeutic effects of arsenic trioxide (As₂O₃) on myeloma cells. Leuk. Res. 25(3):237-239.

Farmer, J.G., and L.R. Johnson. 1990. Assessment of occupational exposure to inorganic arsenic based on urinary concentrations and speciation of arsenic. Br. J. Ind. Med. 47(5):342-348.

Fenaux, P., C. Chomienne, and L. Degos. 2001. All-trans retinoic acid and chemotherapy in the treatment of acute promyelocytic leukemia. Semin. Hemotol. 38(1):13-25.

Feng, Z., Y. Xia, D. Tian, K. Wu, M. Schmitt, R.K. Kwok, and J.L. Mumford. 2001. DNA damage in buccal epithelial cells from individuals chronically exposed to arsenic via drinking water in Inner Mongolia, China. Anticancer Res. 21(1A):51-57.

Foa, V., A. Colombi, M. Maroni, M. Buratti, and G. Calzaferri. 1984. The speciation of the chemical forms of arsenic in the biological monitoring of exposure to inorganic arsenic. Sci. Total Environ. 34(3):241-259.

Forkner, C.E., and T.F. M. Scott. 1931. Arsenic as therapeutic agent in chronic myelogenous leukemia. J.A.M.A 97(1):3-5.

Germolec, D.R., J. Spalding, G.A. Boorman, J.L. Wilmer, T. Yoshida, P.P. Simeonova, A. Bruccoleri, F. Kayama, K. Gaido, R. Tennant, F. Burleson, W. Dong, R.W. Lang, and M.I. Luster. 1997. Arsenic can mediate skin neoplasia by chronic stimulation of keratinocyte-derived growth factors. Mutat. Res. 386(3):209-218.

Germolec, D.R., J. Spalding, H.S. Yu, G.S. Chen, P.P. Simeonova, M.C. Humble, A. Bruccoleri, G.A. Boorman, J.F. Foley, T. Yoshida, and M.I. Luster. 1998. Arsenic enhancement of skin neoplasia by chronic stimulation of growth factors. Am. J. Pathol. 153(6):1775-1785.

Gregus, Z., A. Gyurasics, and I. Csanaky. 2000. Biliary and urinary excretion of inorganic arsenic: monomethylarsonous acid as a major biliary metabolite in rats. Toxicol. Sci. 56(1):18-25.

Hakala, E., and L. Pyy. 1995. Assessment of exposure to inorganic arsenic by determining the arsenic species excreted in urine. Toxicol. Lett. 77(1-3):249-258.

Hamadeh, H.K., M. Vargas, E. Lee, and D.B. Menzel. 1999. Arsenic disrupts cellular levels of p53 and mdm2: a potential mechanism of carcinogenesis. Biochem. Biophys. Res. Commun. 263(2):446-449.

Hayashi, H., M. Kanisawa, K. Yamanaka, T. Ito, N. Udaka, H. Ohji, K. Okudela, S. Okada, and H. Kitamura. 1998. Dimethylarsinic acid, a main metabolite of inorganic arsenics, has tumorigenicity and progression effects in the pulmonary tumors of A/J mice. Cancer Lett. 125(1):83-88.

Healy, S.M., E.A. Casarez, F. Ayala-Fierro, and H. Aposhian. 1998. Enzymatic methylation of arsenic compounds. V. Arsenite methyltransferase activity in tissues of mice. Toxicol. Appl. Pharmacol. 148(1):65-70.

Healy, S.M., E. Wildfang, R.A. Zakharyan, and H.V. Aposhian. 1999. Diversity of inorganic arsenite biotransformation. Biol. Trace Elem. Res. 68(3):249-266.

Holson, J.F., D.G. Stump, K.J. Clevidence, J.F. Knapp, and C.H. Farr. 2000. Evaluation of the prenatal developmental toxicity of orally administered arsenic trioxide in rats. Food Chem. Toxicol. 38(5):459-466.

Hsu, C.H., S.A. Yang, J.Y. Wang, H.S. Yu, and S.R. Lin. 1999. Mutation spectrum of p53 gene in arsenic-related skin cancers from the blackfoot disease endemic area of Taiwan. Br. J. Cancer 80(7):1080-1086.

Hu, Y., L. Su, and E.T. Snow. 1998. Arsenic toxicity is enzyme specific and its affects on ligation are not caused by the direct inhibition of DNA repair enzymes. Mutat. Res. 408(3):203-218.

Huang, R.N., and T.C. Lee. 1996. Cellular uptake of trivalent arsenite and pentavalent arsenate in KB cells cultured in phosphate-free medium. Toxicol. Appl. Pharmacol. 136(2):243-249.

Huang, C., W.Y. Ma, J. Li, A. Goranson, and Z. Dong. 1999a. Requirement of Erk, but not JNK, for arsenite-induced cell transformation. J. Biol. Chem. 274(21):14595-14601.

Huang, C., W.Y. Ma, J. Li, and Z. Dong. 1999b. Arsenic induces apoptosis through a c-Jun NH2-terminal kinase-dependent, p53-independent pathway. Cancer Res. 59(13):3053-3058.

Hughes, M.F., E.M. Kenyon, B.C. Edwards, C.T. Mitchell, and D.J. Thomas. 1999. Strain-dependent disposition of inorganic arsenic in the mouse. Toxicology 137(2):95-108.

Hughes, M.F., and E.M. Kenyon. 1998. Dose-dependent effects on the disposition of monomethylarsonic acid and dimethylarsinic acid in the mouse after intravenous administration. J. Toxicol. Environ. Health A. 53(2):95-112.

Hughes, M.F., and D.J. Thompson. 1996. Subchronic dispositional and toxicological effects of arsenate administered in drinking water to mice. J. Toxicol. Environ. Health 49(2):177-196.

Hughes, M.F., L.M. Del Razo, and E.M. Kenyon. 2000. Dose-dependent effects on tissue distribution and metabolism of dimethylarsinic acid in the mouse after intravenous administration. Toxicology 143(2):155-166.

Hunder, G., J. Schaper, O. Ademuyiwa, and B. Elsenhans. 1999. Species differences in arsenic-mediated renal copper accumulation: a comparison between rats, mice and guinea pigs. Hum. Exp. Toxicol. 18(11):699-705.

Ishitsuka, K., S. Hanada, K. Uozumi, A. Utsunomiya, and T. Arima. 2000. Arsenic trioxide and the growth of human t-cell leukemia virus type i infected t-cell lines. Leuk. Lymphoma 37(5-6):649-655.

Jiang, X.H., B. Chun-Yu Wong, S.T. Yuen, S.H. Jiang, C.H. Cho, K.C. Lai, M.C. Lin, H.F. Kung, and S.K. Lam. 2001. Arsenic trioxide induces apoptosis in human gastric cancer cells through up-regulation of p53 and activation of caspase-3. Int. J. Cancer 91(2):173-179.

Kalman, D.A., J. Hughes, G. van Belle, T. Burbacher, D. Bolgiano, K. Coble, N.K. Mottet, and L. Polissar. 1990. The effect of variable environmental arsenic contamination of urinary concentrations of arsenic species. Environ. Health Perspect. 89:145-151.

Kaltreider, R.C., A.M. Davis, J.P. Lariviere, and J.W. Hamilton. 2001. Arsenic alters the function of the glucocorticoid receptor as a transcription factor. Environ. Health Perspect. 109(3):245-251.

Kaltreider, R.C., C.A. Pesce, M.A. Ihnat, J.P. Lariviere, and J.W. Hamilton. 1999. Differential effects of arsenic (III) and chromium (VI) on nuclear transcription factor binding. Mol. Carcinog. 25(3):219-229.

Kapahi, P., T. Takahashi, G. Natoli, S.R. Adams, Y. Chen, R.Y. Tsien, and M. Karin. 2000. Inhibition of NF-kappa B activation by arsenite through reaction with a critical cysteine in the activation loop of Ikappa B kinase. J. Biol. Chem. 275(46):36062-36066.

Kenyon, E.M., and M.F. Hughes. 2001. A concise review of the toxicity and carcinogenicity of dimethylarsinic acid. Toxicology 160(1-3):227-236.

Kitchin, K.T., L.M. Del Razo, J.L. Brown, W.L. Anderson, and E.M. Kenyon. 1999. An integrated pharmacokinetic and pharmacodynamic study of arsenite action. 1. Heme oxygenase induction in rats. Teratog. Carcinog. Mutagen. 19(6):385-402.

Kuo, T.T., S. Hu, S.K. Lo, and H.L. Chan. 1997. p53 expression and proliferative activity in Bowen's disease with or without chronic arsenic exposure. Hum. Pathol. 28(7):786-790.

Larochette, N., D. Decaudin, E. Jacotot, C. Brenner, I. Marzo, S.A. Susin, N. Zamzami, Z. Xie, J. Reed, and G. Kroemer. 1999. Arsenite induces apoptosis via a direct effect on the mitochondrial permeability transition pore. Exp. Cell Res. 249(2):413-421.

Le, X.C., M. Ma, W.R. Cullen, H.V. Aposhian, X. Lu, and B. Zheng. 2000. Determination of monomethylarsonous acid, a key arsenic methylation intermediate, in human urine. Environ. Health Perspect. 108(11):1015-1018.

Li, D., K. Morimoto, T. Takeshita, and Y. Lu. 2001. Formamidopyrimidine-DNA glycosylase enhances arsenic-induced DNA strand breaks in PHA-stimulated and unstimulated human lymphocytes. Environ. Health Perspect. 109(5):523-526.

Li, Y.M., and J.D. Broome. 1999. Arsenic targets tubulins to induce apoptosis in myeloid leukemia cells. Cancer Res. 59(4):776-780.

Lin, S., W.R. Cullen, and D.J. Thomas. 1999. Methylarsenicals and arsinothiols are potent inhibitors of mouse liver thioredoxin reductase. Chem. Res. Toxicol. 12(10):924-930.

Lin, S., L.M. Del Razo, M. Styblo, C. Wang, W.R. Cullen, and D.J. Thomas. 2001. Arsenicals inhibit thioredoxin reductase in cultured rat hepatocytes. Chem. Res. Toxicol. 14(3):305-311.

Lin, T.H., and Y.L. Huang. 1995. Chemical speciation of arsenic in urine of patients with blackfoot disease. Biol. Trace Elem. Res. 48(3):251-261.

Liou, S.-H., J.C. Lung, Y.H. Chen, T. Yang, L.L. Hsieh, C.J. Chen, and T.N. Wu. 1999. Increased chromosome-type chromosome aberration frequencies as biomarkers of cancer risk in a blackfoot endemic area. Cancer Res. 59(7):1481-1484.

Liu, J., M.B. Kadiiska, Y. Liu, T. Lu, W. Qu, and M.P. Waalkes. 2001a. Stress-related gene expression in mice treated with inorganic arsenicals. Toxicol. Sci. 61(2):314-320.

Liu, J., H. Chen, D.S. Miller, J.E. Saavedra, L.K. Keefer, D.R. Johnson, C.D. Klaassen, and M.P. Waalkes. 2001b. Overexpression of glutathione S-transfer-ase II and multidrug resistance transport proteins is associated with acquired tolerance to inorganic arsenic. Mol. Pharmacol. 60(2):302-309.

Liu, J., Y. Liu, R.A. Goyer, W. Achanzar, and M.P. Waalkes. 2000. Metallothionein-I/II null mice are more sensitive than wild-type mice to the hepatotoxic and nephrotoxic effects of chronic oral or injected inorganic arsenicals. Toxicol. Sci. 55(2):460-467.

Liu, S.X., M. Athar, I. Lippai, C. Waldren, and T.K. Hei. 2001. Induction of oxyradicals by arsenic: implication for mechanism of genotoxicity. Proc. Natl. Acad. Sci. USA 98(4):1643-1648.

Lu, T., J. Liu, E.L. LeCluyse, Y.S. Zhou, M.L. Cheng, and M.P. Waalkes. 2001. Application of cDNA microarray to the study of arsenic-induced liver diseases in the population of Guizhou, China. Toxicol. Sci. 59(1):185-192.

Lynn, S., J.R. Gurr, H.T. Lai, and K.Y. Jan. 2000. NADH oxidase activation is involved in arsenite-induced oxidative DNA damage in human vascular smooth muscle cells. Circ. Res. 86:514-519.

Lynn, S., J.N. Shiung, J.R. Gurr, and K.Y. Jan. 1998. Arsenite stimulates poly(ADP-ribosylation) by generation of nitric oxide. Free Radic. Biol. Med. 24(3):442-449.

Machado, A.F., D.N. Hovland Jr., S. Pilafas, and M.D. Collins. 1999. Teratogenic response to arsenite during neurulation: relative sensitivities of C57BL/6J and SWV/Fnn mice and impact of the splotch allele. Toxicol. Sci. 51(1):98-107.

Maier, A., T.P. Dalton, and A. Puga. 2000. Disruption of dioxin-inducible phase I and phase II gene expression patterns by cadmium, chromium, and arsenic. Mol. Carcinog. 28(4):225-235.

Maki-Paakkanen, J., P. Kurttio, A. Paldy, and J. Pekkanen. 1998. Association between the clastogenic effect in peripheral lymphocytes and human exposure to arsenic through drinking water. Environ. Mol. Mutagen. 32(4):301-313.

Males, R.G., J.C. Nelson, and F.G. Herring. 1998. Vesicular membrane permeability of monomethylarsonic and dimethylarsinic acids. Biophys. Chem. 70(1):75-85.

Mandal, B.K., Y. Ogra, and K.T. Suzuki. 2001. Identification of dimethylarsinous and monomethylarsonous acids in human urine of the arsenic-affected areas in West Bengal, India. Chem. Res. Toxicol. 14(4):371-378.

Mass, M.J., and L. Wang. 1997. Arsenic alters cytosine methylation patters of the promoter of the tumor suppressor gene p53 in human lung cells: a model for a mechanism of carcinogenesis. Mutat. Res. 386(3):263-277.

Mass, M.J., A. Tennant, B.C. Roop, W.R. Cullen, M. Styblo, D.J. Thomas, and A.D. Kligerman. 2001. Methylated trivalent arsenic species are genotoxic. Chem. Res. Toxicol. 14(4):355-361.

Matsui, M., C. Nishigori, S. Toyokuni, J. Takada, M. Akaboshi, M. Ishikawa, S. Imamura, and Y. Miyachi. 1999. The role of oxidative DNA damage in human arsenic carcinogenesis: detection of 8-hydroxy-2'-deoxyguanosine in arsenic-related Bowen's disease. J. Invest. Dermatol. 113(1):26-31.

Menzel, D.B., H.K. Hamadeh, E. Lee, D.M. Meacher, V. Said, R.E. Rasmussen, H. Greene, and R.N. Roth. 1999. Arsenic binding proteins from human lymphoblastoid cells. Toxicol. Lett. 105(2):89-101.

Moore, L.E., A.H. Smith, C. Hopenhayn-Rich, M.L. Biggs, D.A. Kalman, and M.T. Smith. 1997. Micronuclei in exfoliated bladder cells among individuals chronically exposed to arsenic in drinking water. Cancer Epidemiol. Biomarkers Prev. 6(1):31-36.

Morikawa, T., H. Wanibuchi, K. Morimura, M. Ogawa, and S. Fukushima. 2000. Promotion of skin carcinogenesis by dimethylarsinic acid in keratin (K6)/ODC transgenic mice. Jpn. J. Cancer Res. 91(6):579-581.

Murgo, A.J. 2001. Clinical trials of arsenic trioxide in hematologic and solid tumors: overview of the National Cancer Institute Cooperative Research and Development Studies. Oncologist 6(suppl.2):22-28.

Namgung, U., and Z. Xia. 2001. Arsenic induces apoptosis in rat cerebellar neurons via activation of JNK3 and p38 MAP kinases. Toxicol. Appl. Pharmacol. 174(2):130-138.

Ng, J.C. 1999. Speciation, Bioavailability and Toxicology of Arsenic in the Environment. Ph. D. Thesis. University of Queensland, Australia.

NRC (National Research Council). 1999. Arsenic in Drinking Water. Washington, DC: National Academy Press.

Osler, W. 1894. Principles and Practice of Medicine. New York: Appleton.

Park, W.H., J.G. Seol, E.S. Kim, J.M. Hyun, C.W. Jung, C.C. Lee, B.K. Kim, and Y.Y. Lee. 2000. Arsenic trioxide-mediated growth inhibition in MC/CAR myeloma cells via cell cycle arrest in association with induction of cyclin-dependent kinase inhibitor, p21, and apoptosis. Cancer Res. 60(11):3065-3071.

Parrish, A.R., X.H. Zheng, K.D. Turney, H.S. Younis, and A.J. Gandolfi. 1999. Enhanced transcription factor DNA binding and gene expression induced by arsenite or arsenate in renal slices. Toxicol. Sci. 50(1):98-105.

Petrick, J.S., F. Ayala-Fierro, W.R. Cullen, D.E. Carter, and H.V. Aposhian. 2000. Monomethylarsonous acid (MMA^{III}) is more toxic than arsenite in Chang human hepatocytes. Toxicol. Appl. Pharmacol. 163(2):203-207.

Petrick, J.S., B. Jagadish, E.A. Mash, and H.V. Aposhian. 2001. Monometylarsonous acid (MMA III) and arsenite: LD 50 in hamsters and in vitro inhibition of pyruvate dehydrogenase. Chem. Res. Toxicol. 14(6):651-656.

Porter, A.C., G.R. Fanger, and R.R. Vaillancourt. 1999. Signal transduction pathways regulated by arsenate and arsenite. Oncogene 18(54):7794-7802.

Puccetti, E., S. Guller, A. Orleth, N. Bruggenolte, D. Hoelzer, O.G. Ottmann, and M. Ruthardt. 2000. BCR-ABL mediates arsenic trioxide-induced apoptosis independently of its aberrant kinase activity. Cancer Res. 60(13):3409-3413.

Radabaugh, T.R., and H.V. Aposhian. 2000. Enzymatic reduction of arsenic compounds in mammalian systems: reduction of arsenate to arsenite by human liver arsenate reductase. Chem. Res. Toxicol. 13(1):26-30.

Ramirez, P., L.M. Del Razo, and M.E. Gonsebatt. 2000. Arsenite induces DNA-protein crosslinks and cytokeratin expression in the WRL-68 human hepatic cell line. Carcinogenesis 21(4):701-706.

Romach, E.H., C.Q. Zhao, L.M. Del Razo, M.E. Cebrian, and M.P. Waalkes. 2000. Studies on the mechanisms of arsenic-induced self tolerance developed in liver epithelial cells through continuous low-level arsenite exposure. Toxicol. Sci. 54(2):500-508.

Rossman, T.G., and Z. Wang. 1999. Expression cloning for arsenite-resistance resulted in isolation of tumor-suppressor fau cDNA: possible involvement of the ubiquitin system in arsenic carcinogenesis. Carcinogenesis 20(2):311-316.

Roussel, R.R., and A. Barchowsky. 2000. Arsenic inhibits NF-kappaB-mediated gene transcription by blocking IkappaB kinase activity and IkappaBalpha phosphorylation and degradation. Arch. Biochem. Biophys. 377(1):204-212.

Rousselot, P., S. Labaume, J.P. Marolleau, J. Larghero, M.H. Noguera, J.C. Brouet, and J.P. Fermand. 1999. Arsenic trioxide and melarsoprol induce apoptosis in plasma cell lines and in plasma cells from myeloma patients. Cancer Res. 59(5):1041-1048.

Sampayo-Reyes, A., R.A. Zakharyan, S.M. Healy, and H.V. Aposhian. 2000. Monomethylarsonic acid reductase and monomethylarsonous acid in hamster tissue. Chem. Res. Toxicol. 13(11):1181-1186.

Santra, A., J. Das Gupta, B.K. De, B. Roy, and D.N.G. Mazumder. 1999. Hepatic manifestations in chronic arsenic toxicity. Indian J. Gastroenterol. 18(4):152-155.

Santra, A., A. Maiti, S. Das, S. Lahiri, S.K. Charkaborty, and D.N.G. Mazumder. 2000. Hepatic damage caused by chronic arsenic toxicity in experimental animals. J. Toxicol. Clin. Toxicol. 38(4):395-405.

Schroeder, M., and M.J. Mass. 1997. CpG methylation inactivates the transcriptional activity of the promoter of the human p53 tumor suppressor gene. Biochem. Biophys. Res. Commun. 235(2):403-406.

Seol, J.G., W.H. Park, E.S. Kim, C.W. Jung, J.M. Hyun, B.K. Kim, and Y.Y. Lee. 1999. Effect of arsenic trioxide on cell cycle arrest in head and neck cancer cell line PCI-1. Biochem. Biophys. Res. Commun. 265(2):400-404.

Seol, J.G., W.H. Park, E.S. Kim, C.W. Jung, J.M. Hyun, Y.Y. Lee, and B.K. Kim. 2001. Potential role of caspase-3 and -9 in arsenic trioxide-mediated apoptosis in PCI-1 head and neck cancer cells. Int. J. Oncol. 18(2):249-255.

Shen, Z.Y., L.J. Tan, W.J. Cai, J. Shen, C. Chen, X.M. Tang, and M.H. Zheng. 1999. Arsenic trioxide induces apoptosis of oesophageal carcinoma in vitro. Int. J. Mol. Med. 4(1):33-37.

Simeonova, P.P., S. Wang, W. Toriuma, V. Kommineni, J. Matheson, N. Unimye, F. Kayama, D. Harki, M. Ding, V. Vallyathan, and M.I. Luster. 2000. Arsenic mediates cell proliferation and gene expression in the bladder epithelium: association with activating protein-1 transactivation. Cancer Res. 60(13):3445-3453.

Simeonova, P.P., S. Wang, M.L. Kashon, C. Kommineni, E. Crecelius, and M.I. Luster. 2001. Quantitative relationship between arsenic exposure and AP-1 activity in mouse urinary bladder epithelium. Toxicol. Sci. 60(2):279-284.

Smith,T., E.A.Crecelius, and J.C. Reading. 1977. Airborne arsenic exposure and excretion of methylated arsenic compounds. Environ. Health Perspect. 19:89-93.

Snow, E.T., Y. Hu, C.C. Yan, and S. Chouchane. 1999. Modulation of DNA repair and glutathione levels in human keratinocytes by micromolar arsenite. Pp. 243-251 in Arsenic Exposure and Health Effects, W.R. Chappell, C.O. Abernathy, and R.L. Calderon, eds. Oxford: Elsevier.

Snow, E.T., M. Schuliga, S. Chouchane, and Y.Hu. In press. Sub-toxic arsenite induces a multi-component protective response against oxidative stress in human cells. In Arsenic Exposure and Health Effects. Proceedings of the 4th International SEGH Conference on Arsenic Exposure and Health, June 18-22, 2000., W.R. Chappell, C.O. Abernathy, and R.L. Calderon, eds. Oxford: Elsevier

Soignet, S., E. Calleja, N.-K. Cheung, S. Pezzulli, P. Vongphrachanh, D. Spriggs, and R.P. Warrell. 1999. A Phase 1 Study of Arsenic Trioxide in Patients with Solid Tumors, Memorial Sloan-Kettering Cancer Center, New York, NY. Program/ Proceedings Abstracts for 35th Annual Meeting of the American Society of Clinical Oncology, Atlanta, GA, May 15-18, 1999. Vol. 18. Clinical Pharmacology Abstract no. 878. [Online]. Available: http://www.asco.org/ prof/me/html/ 99abstracts/ m_toc.htm [August 24, 2001].

Styblo, M., and D.J. Thomas. 1997. Binding of arsenicals to proteins in an in vitro methylation system. Toxicol. Appl. Pharmacol. 147(1):1-8.

Styblo, M., L.M. Del Razo, E.L. LeCluyse, G.A. Hamilton, C. Wang, W.R. Cullen, and D.J. Thomas. 1999a. Metabolism of arsenic in primary cultures of human and rat hepatocytes. Chem. Res. Toxicol. 12(7):560-565.

Styblo, M., L.M. Del Razo, L. Vega, D.R. Germolec, E.L. LeCluyse, G.A. Hamilton, W. Reed, C. Wang, W.R. Cullen, and D.J. Thomas. 2000. Comparative toxicity of trivalent and pentavalent inorganic and methylated arsenicals in rat and human cells. Arch. Toxicol. 74(6):289-299.

Styblo, M., S.V. Serves, W.R. Cullen, and D.J. Thomas. 1997. Comparative inhibition of yeast glutathione reductase by arsenicals and arsenothiols. Chem. Res. Toxicol. 10(1):27-33.

Styblo, M., L. Vega, D.R. Germolec, M.I. Luster, L.M. Del Razo, C. Wang, W.R. Cullen, and D.J. Thomas. 1999b. Metabolism and toxicity of arsenicals in cultured cells. Pp. 311-323 in Arsenic Exposure and Health Effects, W.R. Chappell, C.O. Abernathy, and R.L. Calderon, eds. Oxford: Elsevier.

Tatum, F.M., and R.D. Hood. 1999. Arsenite uptake and metabolism by rat hepatocyte primary cultures in comparison with kidney- and hepatocyte-derived rat cell lines. Toxicol. Sci. 52(1):20-25.

Thompson, D.J. 1993. A chemical hypothesis for arsenic methylation in mammals. Chem. Biol. Interact. 88(2-3):89-114.

Trouba, K.J., E.M. Wauson, and R.L. Vorce. 2000a. Sodium arsenite inhibits terminal differentiation of murine C3H 10T1/2 preadipocytes. Toxicol. Appl. Pharmacol. 168(1): 25-35.

Trouba, K.J., E.M. Wauson, and R.L. Vorce. 2000b. Sodium arsenite-induced dysregulation of proteins involved in proliferative signaling. Toxicol. Appl. Pharmacol. 164(2):161-170.

Tully, D.B., B.J. Collins, J.D. Overstreet, C.S. Smith, G.E. Dinse, M.M. Mumtaz, and R.E. Chapin. 2000. Effects of arsenic, cadmium, chromium and lead on gene expression regulated by a battery of 13 different promoters in recomibnant HepG2 cells. Toxicol. Appl. Pharmacol. 168(2):79-90.

Uslu, R., U.A. Sanli, C. Sezgin, B. Karabulut, E. Terzioglu, S.B. Omay, and E. Goker. 2000. Arsenic trioxide-mediated cytotoxicity and apoptosis in prostate and ovarian carcinoma cell lines. Clin. Cancer Res. 6(12):4957-4964.

Vahter, M. 1999a. Methylation of inorganic arsenic in different mammalian species and population groups. Sci. Prog. 82(Pt.1):69-88.

Vahter, M. 1999b. Variation in human metabolism of arsenic. Pp. 267-279 in Arsenic Exposure and Health Effects, W.R. Chappell, C.O. Abernathy, and R.L. Calderon, eds. Oxford: Elsevier.

Vega, L., P. Ostrosky-Wegman, T.I. Fortoul, C. Diaz, V. Madrid, and R. Saavedra. 1999. Sodium arsenite reduces proliferation of human activated T-cells by inhibition of the secretion of interleukin-2. Immunopharmacol. Immunotoxicol. 21(2):203-220.

Vega, L., M. Styblo, R. Patterson, W. Cullen, C. Wang, and D. Germolec. 2001. Differential effects of trivalent and pentavalent arsenicals on cell proliferation and cytokine secretion in normal human epidermal keratinocytes. Toxicol. Appl. Pharmacol. 172(3):225-232.

Vogt, B.L., and T.G. Rossman. 2001. Effects of arsenite on p53, p21 and cyclin D expression in normal human fibroblasts - a possible mechanism for arsenite's comutagenicity. Mutat. Res. 478(1-2):159-168.

Waalkes, M.P., L.K. Keefer, and B.A. Diwan. 2000. Induction of proliferative lesions of the uterus, testes, and liver in swiss mice given repeated injections of sodium arsenate: possible estrogenic mode of action. Toxicol. Appl. Pharmacol. 166(1):24-35.

Wei, M., H. Wanibuchi, S. Yamamoto, W. Li, and S. Fukushima. 1999. Urinary bladder carcinogenicity of dimethylarsinic acid in male F344 rats. Carcinogenesis 20(9):1873-1876.

Wildfang, E., R.A. Zakharyan, and H.V. Aposhian. 1998. Enzymatic methylation of arsenic compounds. VI. Characterization of hamster liver arsenite and methylarsonic acid methyltransferase activities in vitro. Toxicol. Appl. Pharmacol. 152(2):366-375.

Yamamura,Y., and H. Yamauchi. 1980. Arsenic metabolites in hair, blood and urine in workers exposed to arsenic trioxide. Ind. Health 18(4):203-210.

Yamanaka, K., K. Katsumata, K. Ikuma, A. Hasegawa, M. Nakano, and S. Okada. 2000. The role of orally administered dimethylarsinic acid, a main metabolite of inorganic arsenics, in the promotion and progression of UVB-induced skin tumorigenesis in hairless mice. Cancer Lett. 152(1):79-85.

Yang, C.H., M.L. Kuo, J.C. Chen, and Y.C. Chen. 1999. Arsenic trioxide sensitivity is associated with low level of glutathione in cancer cells. Br. J. Cancer 81(5):796-799.

Yih, L.H., and T.C. Lee. 1999. Effects of exposure protocols on induction of kinetochore-plus and -minus micronuclei by arsenite in diploid human fibroblasts. Mutat. Res. 440(1):75-82.

Yoshida, K., Y. Inoue, K. Kuroda, H. Chen, H. Wanibuchi, S. Fukushima, and G. Endo. 1998. Urinary excretion of arsenic metabolites after long-term oral administration of various arsenic compounds to rats. J. Toxicol. Environ. Health 54(3):179-192.

Zakharyan, R.A., and H.V. Aposhian. 1999a. Arsenite methylation by methylvitamin B_{12} and glutathione does not require an enzyme. Toxicol. Appl. Pharmacol. 154(3):287-291.

Zakharyan, R.A., and H.V. Aposhian. 1999b. Enzymatic reduction of arsenic compounds in mammalian systems: the rate-limiting enzyme of rabbit liver arsenic biotransformation is MMA^v reductase. Chem. Res. Toxicol. 12(12):1278-1283.

Zakharyan, R.A., F. Ayala-Fierro, W.R. Cullen, D.M. Carter, and H.V. Aposhian. 1999. Enzymatic methylation of arsenic compounds. VII. Monomethylarsonous acid (MMA^{III}) is the substrate for MMA methyltransferase of rabbit liver and human hepatocytes. Toxicol. Appl. Pharmacol. 158(1):9-15.

Zakharyan, R.A., A. Sampayo-Reyes, S.M. Healy, G. Tsaprailis, P.G. Board, D.C. Liebler, and H.V. Aposhian. 2001. Human Monomethylarsonic Acid (MMA(V)) reductase is a member of the glutathione-S-transferase superfamily. Chem. Res. Toxicol. 14(8):1051-1057.

Zhang, T.C., E.H. Cao, J.F. Li, W. Ma, and J.F. Qin. 1999. Induction of apoptosis and inhibition of human gastric cancer MGC-803 cell growth by arsenic trioxide. Eur. J. Cancer 35(8):1258-1263.

Zhang, W., K. Ohnishi, K. Shigeno, S. Fujisawa, K. Naito, S. Nakamura, K. Takeshita, A. Takeshita, and R. Ohno. 1998. The induction of apoptosis and cell cycle arrest by arsenic trioxide in lymphoid neoplasms. Leukemia 12(9):1383-1391.

Zhong, C.X., and M.J. Mass. 2001. Both hypomethylation and hypermethylation of DNA associated with arsenite exposure in cultures of human cells identified by methylation-sensitive. Toxicol. Lett. 122(3):223-234.

4

Variability and Uncertainty

VARIABILITY VERSUS UNCERTAINTY

Quantitative health risk assessments rely on mathematical models to combine data and assumptions with the goal of providing useful information to the decision-maker about health risks. In the context of developing a dose-response relationship, risk assessors evaluate the weight of the scientific evidence of an effect. Given sufficient evidence, they combine data from multiple studies in the same species and from epidemiological and animal studies to estimate the health risks. These risks are estimated in the dose range for which observations have been made and are then extrapolated from these dose ranges to lower doses that might be of concern for regulation. *Science and Judgment in Risk Assessment* (NRC 1994) emphasized the important distinctions between variability and uncertainty in risks, highlighting their very different ramifications for risk management in the following statement:

> Uncertainty forces decision-makers to judge how *probable* it is that risks will be overestimated or underestimated for every member of the exposed population, whereas variability forces them to cope with the *certainty* that different individuals will be subjected to risks both above and below any reference point one chooses (NRC 1994, p. 237).

Uncertainty refers to a lack of knowledge in the underlying science. Variability in a risk assessment means that some individuals in a population have more or less risk than others because of differences in exposure, dose-response relationship, or both. *Arsenic in Drinking Water* (NRC 1999) discussed the variability and uncertainty associated with an arsenic risk assessment. This chapter focuses on data that have become available since that report. These data provide further insight into the variability and uncertainty in the risk assessment and allow a better characterization of how they affect the overall characterization of risk. This chapter is divided into two sections, one that primarily discusses variability and one that primarily discusses uncertainty. It is important, however, to recognize that uncertainty might exist about variable quantities and that variability and uncertainty are best defined and understood within a decision-making context (NRC 1994; Thompson and Graham 1996).

VARIABILITY AND UNCERTAINTIES
DISCUSSED IN THE 1999 REPORT

In its discussion of variability in an arsenic risk assessment, the 1999 Subcommittee on Arsenic in Drinking Water concluded that characteristics of different individuals could contribute to variability in risk estimates. Human susceptibility to inorganic arsenic could vary because of differences in genetics, metabolism, diet, health status, and sex.

In the 1999 NRC report, the subcommittee presented risk estimates that were calculated using different statistical models, which provide some indication of the impact of the model on the estimates, and concluded that the statistical model used to analyze the data can have a major impact on the estimated cancer risks at low-dose exposures. The 1999 NRC report quantified some of the impacts of different dose-response model choices, but did not quantify the impacts of variability in the population of the risk estimates.

In most epidemiological studies of arsenic, biomarkers of individual arsenic exposure have not been used; therefore, assumptions must be made about the amount and source of water consumed to estimate exposures. Those assumptions add to the uncertainty in the risk assessment. There is the potential for bias in the results of the Taiwanese studies (Chen et al. 1985, 1988, 1992; Wu et al. 1989) because of the uncertainty in exposure estimates. However, similar associations between arsenic exposure and cancers (lung and bladder) were seen in studies in Chile (Smith et al. 1998) and Argentina (Hopenhayn-Rich et al. 1998); therefore, the uncertainty in the Taiwanese data, and the subsequent risk assessment using those data, is decreased (NRC 1999). Additional uncertainty comes from the lack of knowledge of the precise intake of arsenic from food in both the U.S. and Taiwanese populations. Differences in the intake between the populations could affect the assessment of risk from exposure to arsenic in Taiwan and the extrapolation of that risk to the U.S. population. The lack of knowledge regarding population differences adds to the uncertainty in the risk estimates.

SOURCES OF VARIABILITY

Exposure

Different individuals or subpopulations are exposed to different amounts of arsenic, leading to variability in the risk estimates that could be calculated for a given population. Many factors can affect the amount of arsenic to which different individuals are exposed through drinking water, including differences in the amount of arsenic present in drinking water, the amount of water that individuals consume, and physical factors, such as sex, age, and body weight. Exposure to arsenic through sources other than drinking water can also lead to variability in risk estimates. Environmental arsenic concentrations affect exposures that do not come from drinking water, and those concentrations vary among locations because of different natural background levels and local anthropogenic sources. In addition, different foods contain varying concentrations of arsenic (IOM 2001); therefore, an individual's food preferences will affect arsenic intake. Although this update focuses on risk from exposure to arsenic in drinking water, other factors that affect arsenic exposure from all sources are important to consider when interpreting the dose-response modeling based on use of available epidemiological data. This section discusses those sources of variability in exposure estimates and their potential impacts on an arsenic risk assessment. Many of those sources are further explored and discussed in Chapter 5.

Per Capita Water Consumption

Water Consumption in the U.S. Population

One reason a subpopulation might be at higher risk for the adverse effects of arsenic in drinking water is that they have a higher than average water consumption per kilogram of body weight. The previous subcommittee (NRC 1999) and Morales et al. (2000) estimated lifetime cancer risks using EPA's default water consumption value of 2 L/day, based on a report by Ershow et al. (1991) and assuming a 70-kg person in the United States. Use of these single default values, however, does not address variability in the U.S. population with respect to different water consumption patterns or body weights.

In its final rule, EPA (2001) addressed that variability in its risk assessment (EPA 2000a), and in support of its risk assessment, released updated data on per capita water ingestion in the United States that provided information about variability in water consumption (EPA 2000b). Those new data allow determination of the impact of differential drinking-water rates in the U.S. population on the risk assessment. However, limited information exists on changes that might occur in drinking-water consumption patterns over time.

The new EPA (2000b) analysis of drinking-water rates is based on data collected by the U.S. Department of Agriculture (USDA) as part of its periodic dietary survey, the Continuing Survey of Food Intakes by Individuals (CSFII). The CSFII measures intake of food and beverages for 2 nonconsecutive days and is designed to be a representative probability sample of the entire U.S. population. Recognizing the lack of information about some particular groups in the population, the most recent CSFII collected in 1994-1996 included population subsets for several different age and gender categories and for pregnant, lactating, and childbearing women. Although the CSFII has been conducted several times, the questions asked of the participants have changed each time, making longitudinal analyses of the studies complicated.

In the 1994-1996 collection period, which surveyed over 15,000 Americans, the questions related to drinking water asked the participants specifically about the amount of water consumed from various sources, including community water (tap water from the community water supply), bottled water, and other sources (including private wells, rain cisterns, and household and public springs). The report also adds these amounts to provide estimates of total water consumed (from all sources combined). In addition to using the drinking-water data to estimate arsenic exposure, EPA (2000b) also used the survey data on foods, combined with recipe and nutrient data, to estimate indirect water consumption by each participant over the 2 nonconsecutive days. Indirect water use includes water used in the final preparation of foods and beverages at home and water used by food-service establishments, including restaurants and cafeterias, but it excludes commercial and biological water in foods. EPA reported the results of its water analysis in a variety of ways, including by individual sex and combined, by broad and fine age categories, by water-source type, and by type of consumption (direct, indirect, and total).

For many of these permutations, EPA (2000b) provided the raw data that could be used to fit empirical distributions to the data set. Table 4-1 provides an example of the data provided, and Figure 4-1 shows the plot of the distribution.

The data clearly show variability in water consumption with age, with the highest daily total water ingestion rates per unit of body weight occurring in infants and young children. In its risk assessment, EPA took the reported ingestion rates (given in milliliters per day) from an earlier draft of the EPA (2000b) report. EPA developed distributions of body weight from national census data to get the distribution of drinking-water rates on the basis of per unit of body weight. Ultimately, the EPA (2000b) report on drinking-water

TABLE 4-1 Example of the Distribution of the Water Intakes Represented As Selected Percentiles Reported by EPA (2000b) for Both Sexes, Total Water (from All Sources), and Direct and Indirect Pathways Averaged over 2 Nonconsecutive Days (mL/kg/day)

Fine-age Category (year)	Mean	Percentile								
		1%	5%	10%	25%	50%	75%	90%	95%	99%
<0.5	92	2[a]	7[a]	14	31	87	139	169	196[a]	239[a]
0.5-0.9	65	2[a]	6[a]	11	26	58	88	120	164[a]	185[a]
1-3	31	1	4	7	15	26	40	60	74	118[a]
4-6	27	1[a]	5	8	14	23	36	51	68	97[a]
7-10	20	1[a]	4	6	10	17	26	36	44	70[a]
11-14	16	1[a]	3	4	8	14	21	33	40	60[a]
15-19	15		2	4	7	12	19	29	38	66[a]
20-24	18		2	4	8	14	22	34	44	86[a]
25-54	20	1	4	6	11	17	26	37	46	69
55-64	20	2	6	8	12	18	26	35	42	59
65+	21	4	7	9	13	19	27	34	39	54
All ages	21	1	4	6	11	17	26	38	50	87

[a] Sample size does not meet the minimum reporting requirements.
Source: EPA (2000b, p. IV-19).

rates also presented the distribution in per unit of body weight, and the distributions were very similar. In its analysis, EPA computed lifetime average water consumption by randomly selecting from (1) the water consumption and body-weight distributions for each sex and age category, and (2) randomly selecting a sex for the simulated individual, then multiplying the values drawn by the number of years the individual male or female spent in each age category, and finally summing over the age categories to simulate an entire lifetime. This method did not make any assumptions about consistency (or correlation) in individual body weights or water-consumption rates over time.

The very short sample time of the water intake survey deserves additional discussion. Although the data on water-consumption rate provide a large amount of information about water ingestion for a very short time (2 days), several important uncertainties arise when using these short-term data to represent the long-term average. To compute the average lifetime ingestion rate from those data, the changes in individual drinking-water behaviors over time must be assumed. It must be also assumed that a random cross-sectional sample for 2 nonconsecutive days accurately estimates water consumption over a lifetime. The later assumption might be valid if the 2 days are representative, because it might be expected that intraindividual variability in water consumption over time will be lower than interindividual variability over time (i.e., people who drink a lot of water 1 year will probably drink a lot of water other years). However, the data are lacking to test this assumption, and the extrapolation from a 2-day sample to a longer time period might be misleading if some people participate in the survey during a hot summer week and others respond in the winter, neither giving an estimate of typical consumption. In general, some regression to the mean should be expected as a longer period of time is sampled. Biases associated with the survey and participant compliance also add to the uncertainties in the data. Despite those caveats, EPA's data are useful for risk analysis and for characterizing the variability in population drinking-water rates per unit of body weight. Those data are used in the risk calculations done by the subcommittee to determine the sensitivity of the risk estimates to drinking-water consumption values (presented in Chapter 5).

Water Consumption in the Taiwanese Study Population

In its earlier risk assessment, EPA (1988) did not consider variability in water consumption per unit of body weight in the study populations when interpreting the epidemiological data on arsenic carcinogenicity in Taiwan or

(a)

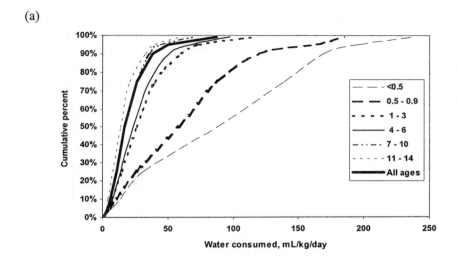

(b)

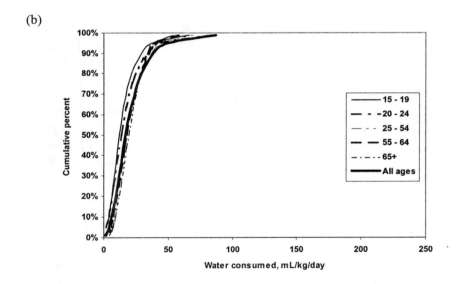

FIGURE 4-1 Cumulative distributions of water intake data by age. (a) Cumulative distributions at ages less than 0.5, 0.5-0.9, 1-3, 4-6, 7-10, 11-14, and all ages. (b) Cumulative distributions at ages 15-19, 20-24, 25-54, 55-64, 65 and above, and all ages.

the dose-response relationship; the previous subcommittee also did not quantify that variability (NRC 1999). EPA (1988) and NRC (1999) assumed that a typical Taiwanese male and female weigh 55 kg and 50 kg, respectively, and drink 3.5 L/day and 2.0 L/day, respectively. Those values correspond to a daily per kilogram intake of water of approximately 60-40 mL/kg/day, much more than the 21-28 mL/kg/day that Americans are estimated to consume. The assumptions suggest considerable differences between Taiwanese and American populations, and the assumptions about Taiwanese water intake have been questioned previously (Mushak and Crocetti 1995). Unfortunately, uncertainty remains about how much water the Taiwanese drink, and no appropriate data on the distribution of water consumption in the Taiwanese study populations are available at this time. When such data become available, they could be used to better estimate exposures in the Taiwanese study populations, reducing the uncertainty in the risk assessment. However, it seems likely that the water consumption pattern of the people in southwestern Taiwan, where many of the previous epidemiological studies were carried out, has changed with time as the socioeconomic situation has improved. Therefore, information on current water intakes might not be useful for reevaluation of the situation 20-40 years ago. In the absence of reliable data on water consumption in the Taiwanese study populations, the sensitivity of the risk estimates to those assumptions should be assessed to quantify some of the uncertainty in the risk assessment. Some analyses of the sensitivity of the risk estimates to different Taiwanese water-intake assumptions are presented in Chapter 5.

Given that there is variability in daily drinking-water ingestion rates, which can be partially characterized based on available data, the subcommittee believes that the EPA risk assessment should explicitly address the impacts of this variability. In particular, the subcommittee notes that the recent approach of EPA (2001) to account for variability in quantity of water consumed by the United States population without simultaneously accounting for variability in water consumption among the Taiwanese would have the result of underestimating the upper bound of risk attributable to arsenic in drinking water in the quantitative risk assessment. It is reasonable to assume that the Taiwanese had a variable intake of water as well, and biostatistical analysis should consider that the southwestern Taiwanese cohort consisted of a significant proportion of males and females whose direct water consumption was less than 3.5 and 2.0 L/day, respectively.

Consideration of Infants and Children

Within a given population, certain individuals or subpopulations can be more susceptible to the toxic effects of a substance, either because they have a higher exposure to the substance or because they are intrinsically more susceptible to it. Infants and children are often considered more susceptible to the adverse effects of toxic substances than adults (NRC 1993), but susceptibility must be assessed on a case-by-case basis (ILSI 1992).

Susceptibility from Exposure Differences

EPA (2000b) noted that "when considering water-ingestion rates in units of milliliters per kilogram of body weight per day, this analysis shows that the mean per capita ingestion rates for babies younger than one year are estimated to be three to four times higher than the mean rates for the population as a whole." That increased exposure (on a body-weight basis) indicates that infants and young children might be at increased risk.[1] Therefore, the subcommittee considered whether there is any evidence that infants and children are more susceptible than adults to the toxic effects of arsenic in drinking water. Calderon et al. (1999) investigated the excretion of arsenic in urine. They found increased concentrations of arsenic metabolites (measured as total arsenic) in individuals less than 18 years of age compared with individuals greater than 18 years of age and suggested that the difference might be due to differences in water consumption. Although these data might support the hypothesis that children are at higher risk to the effects of arsenic in drinking water, the impact of the increased exposure in infants and young children on the risk assessment depends on what pattern of exposure correlates with the effect and whether the increased exposure is intrinsically accounted for in the design of the epidemiological studies used in the risk assessment. Those issues are discussed further in the section Dose Metrics and Model Uncertainty.

---

[1]It should be noted that the variation in water intake of infants is dependent on the frequency of breast feeding.

Intrinsic Susceptibility

It is also important to consider whether infants and children are intrinsically more susceptible to arsenic toxicity. Differences in the metabolism of arsenic in infants, children, and adults could make one of those subpopulations more susceptible to the toxic effects of arsenic. Previous studies indicate that young individuals might have a lower rate of methylation of inorganic arsenic than adults (NRC 1999). More recently, indigenous Andean children have been found to have about 45% of their urinary arsenic as dimethylarsinic acid (DMA) (Concha et al. 1998). In a study of people (2-83 years of age) exposed to arsenic in drinking water in Finland, the percentage of arsenic in urine as DMA increased with age (from 59% in people less than 30 years of age to 74% in people more than 50 years of age) (Kurttio et al. 1998). Similar results were seen in a study of arsenic-exposed people in northeastern Taiwan (Hsu et al. 1997). In contrast, a recent study from Mexico showed that children in Region Lagunera had normal fractions of DMA in urine (an average of about 70%; Del Razo et al. 1999). Therefore, the data on variation in arsenic methylation with age are not consistent, and it is not known how such differences in methylation would affect arsenic toxicity. As discussed in Chapter 3, the methylation of inorganic arsenic can affect its toxicity, but that methylation is no longer thought to be entirely a detoxification process.

Variability in Arsenic Metabolism

Some subsections or subpopulations of a given population could be more or less susceptible to arsenic toxicity because of differences in arsenic metabolism. In particular, there are marked differences in arsenic methylation among mammalian species, population groups, and individuals. As discussed in Chapter 3, the valence state and extent of methylation of arsenic can affect its toxicity. The main products in the methylation of inorganic arsenic, pentavalent monomethylarsonic acid (MMA^V) and dimethylarsinic acid (DMA^V), are readily excreted in urine. Therefore, more efficient methylation, especially to DMA^V, means faster overall excretion of arsenic (Vahter 1999a). The methylation of arsenic, however, is not entirely a detoxifying process. The initial reduction of pentavalent inorganic arsenic (As^V) to trivalent inorganic arsenic (As^{III}), as well as the formation and distribution of methylated trivalent arsenic metabolites in tissues, results in increased toxicity. MMA^{III} is highly reactive with cellular constituents, especially sulfhydryl groups, and, in general, exhibits a greater toxicity than inorganic As^{III} (see Chapter 3). Variations in the metabolism of arsenic, therefore, are likely to be associated with variations in susceptibility to arsenic. The extent to which such metabolic variations might affect an individual's susceptibility to cancer or systemic toxicity with arsenic exposure is an important uncertainty.

Because methylated forms of arsenic are readily excreted in urine, evaluation of arsenic methylation efficiency is generally based on the relative amounts of the different metabolites detected in urine (Vahter 1999a). Currently, however, there are no data on the association between urinary concentrations of MMA^{III} or DMA^{III} and their formation and concentrations in tissues. Those metabolites are highly reactive and retained intracellularly, whereas DMA^{V} is the main form of arsenic excreted from the cells (Styblo et al. 2000) (see Chapter 3). Therefore, the concentrations of MMA^{III} and DMA^{III} in the urine are less likely to reflect the formation of those metabolites in tissues. However, MMA^{III} has been detected in human urine, although the site of reduction of MMA^{V} to MMA^{III} is not known. Because MMA^{III} is produced from MMA^{V}, it seems probable that the tissue concentrations of MMA^{III} would increase with the total production of MMA and, therefore, with increasing total MMA in urine.

Genetic Polymorphisms Related to Arsenic Metabolism

Glutathione (GSH) S-transferases (GST) constitute a large family of detoxifying enzymes that catalyze the conjugation of reduced GSH with a wide range of compounds. Because GSH levels, which are essential for the reduction and methylation of arsenic, are influenced by GSTs, the latter have been considered as possible candidates for influencing the metabolism and the toxicity of arsenic. A study from the area in northeastern Taiwan where arsenic is endemic showed that subjects with the null genotype of GST M1 had a slightly increased percentage of inorganic arsenic in urine (Chiou et al. 1997). In addition, there was a tendency for an increased percentage of DMA in urine of subjects with the null genotype of GST T1. Whether those genotypes might influence susceptibility to arsenic is not known. As discussed in Chapter 3, recent studies indicate that human MMA^{V} reductase is identical to human glutathione-s-transferase omega class 1-1 (hGSTO 1-1). Therefore, polymorphisms in that GST also might affect both cellular protective mechanisms and arsenic metabolism. hGSTO 1-1 has not been extensively studied (see Chapter 3), and it is not known if polymorphisms in that gene occur in human populations.

A polymorphism in the *N*-acetyltransferase-2 (NAT2*) gene can affect susceptibility to bladder cancer. Previous studies have demonstrated that individuals who are slow acetylators as a result of a polymorphism in that gene are at greater risk for developing bladder cancer following a chemical exposure (to arylamines). Therefore, Su et al. (1998) investigated whether that polymorphism also makes individuals susceptible to arsenic-induced bladder cancer. The NAT2*-related slow *N*-acetylation polymorphism was not associated with increased risk for bladder cancer among arsenic-exposed people in the area of southwestern Taiwan where blackfoot-disease is endemic. An association was seen, however, in an area of Taiwan without increased arsenic in drinking water (Su et al. 1998).

Population Variations in Arsenic Metabolism

The average relative distribution of arsenic metabolites in the urine of various population groups exposed to arsenic seems to be fairly consistent: 10-30% inorganic arsenic, 10-20% MMA (generally lower than the percentage of inorganic arsenic), and 60-70% DMA (for review, see Vahter 1999b). There are, however, exceptions. Indigenous people in the north of Argentina and Chile often excrete very little MMA in urine, and in some cases, only a few percent of the arsenic is MMA (NRC 1999). In a recent study in Region Lagunera, Mexico, Del Razo et al. (1999) found that in children who consumed 100-1,250 µg arsenic/L in drinking water, less than 10% of urinary arsenic was MMA in about 40% of the children, and less than 5% was MMA in about 20% of the children. In contrast, studies from southwestern Taiwan consistently show a higher percentage of MMA (usually 20-30%) than inorganic arsenic in urine. In people living in northeastern Taiwan, where arsenic concentrations in drinking water might reach 3,000 µg/L, 27% of urinary arsenic was MMA (Chiou et al. 1997); inorganic arsenic constituted only 12% of total urinary arsenic metabolites. Because high arsenic intake might inhibit the second methylation step (i.e., from MMA to DMA), increased concentrations of MMA might be due in part to the high exposure concentrations. However, the subset of Taiwanese individuals who consumed water with arsenic at less than 50 µg/L, a concentration at which the second methylation step is unlikely to be inhibited, also had, on average, 31.5% and 12% of their urinary arsenic as MMA and inorganic arsenic, respectively. In another study from southwestern Taiwan, where people no longer drink the artesian water with high arsenic concentrations, 21% of the average urinary arsenic was MMA (Hsueh et al. 1998). In skin-cancer patients from the same area, 31% of urinary arsenic was MMA, and 11% was inorganic arsenic; in controls, 23% was MMA, and 11% was inorganic arsenic. Thus, the average percentage of inorganic arsenic in total urinary arsenic was less than half of the percentage of MMA. Total urinary arsenic was less than 100 µg/L, which is not likely to markedly influence the percentage of MMA.

Among the populations studied, less variation appears to occur in urinary DMA than in MMA (Figure 4-2). People exposed to arsenic concentrations that would not inhibit methylation generally have more than 60% of their urinary arsenic as DMA. There are marked differences in the percentage of MMA in urine among mammalian species (see Chapter 3).

The observed variations in the pattern of arsenic metabolites in urine might indicate a genetic polymorphism in the regulation of enzymes responsible for arsenic methylation. In particular, the reduction of MMA^V to MMA^{III} and subsequent methylation of MMA^{III} to DMA^{III} might be slow in certain people, resulting in high MMA concentrations in urine. The high urinary MMA concentrations might also be the result of low binding affinity of MMA to a carrier protein or more efficient mechanisms for excreting MMA from the cell.

A few studies published to date on MMA^{III} and DMA^{III} indicate variations in the excretion of these arsenic metabolites in urine. People exposed to arsenic in drinking water in Romania at 28-161 µg/L were found to excrete MMA^{III} at 5-7 µg/L, on average, independent of the exposure concentration; in those individuals with very low exposures (2.8 µg/L), MMA^{III} was not detectable in urine because of the low concentrations (Aposhian et al. 2000b). Four groups in Bangladesh, where average arsenic concentrations in water are between 33 and 248 µg/L, showed increasing urinary MMA^{III} and DMA^{III} concentrations with increasing exposures (Mandal et al. 2001). However, only 42% of those exposed had urinary MMA^{III}, and 72% had DMA^{III}. The average concentrations of MMA^{III} for the groups ranged from 3 to 30 µg/L, and those of DMA^{III} ranged from 8 to 64 µg/L. In contrast, people in Inner Mongolia, China, exposed to arsenic at about 500 µg/L of drinking water (Aposhian et al. 2000a), and people in northern Chile, exposed at about 600 µg/L (Aposhian et al. 1997), excreted MMA^{III} in urine only after treatment with 2,3-dimercaptopropane sulfonic acid (DMPS). It should be noted that trivalent arsenic metabolites are easily oxidized following the collection of

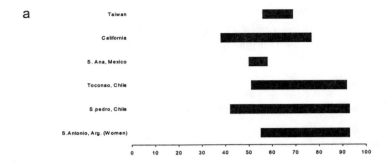

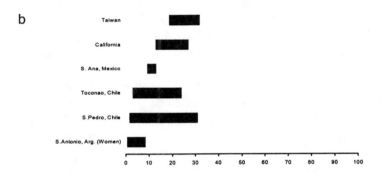

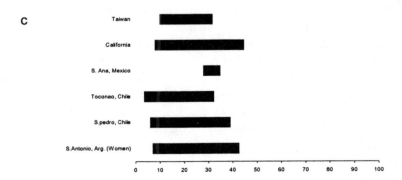

FIGURE 4-2 Population variations in the percentage of urinary as various arsenic metabolites. (a) Percentage as DMA, (b) Percentage as MMA, and (c) Percentage as inorganic arsenic. Source: Modified from Vahter 2000.

urine samples (Le et al. 2000), and the oxidation might have resulted in underestimation of the amounts of MMA[III] and DMA[III] in urine. That might explain some of the differences seen between studies. Therefore, firm conclusions on population variations must await further investigation of the effect of sampling, storage, and analyses on urinary concentrations of MMA[III] and DMA[III].

Interindividual Variation in Arsenic Metabolism

There is considerable interindividual variation in arsenic methylation in humans, possibly due to variations in the activities of the enzymes involved in arsenic methylation. In vitro studies of the methylation of inorganic arsenic in human hepatocytes obtained from four donors showed variation in arsenic methylation between 3 and 6 picomoles (pmol) of arsenite per 10^6 cells per hour (Styblo et al. 1999). There is also wide interindividual variation in the relative amounts of MMA and DMA in urine (Vahter 1999b, 2000). The distribution pattern for the arsenic metabolites in the urine of individual Argentinian women exposed to arsenic in drinking water was remarkably stable over a period of about a week, indicating that an individual's methylation of inorganic arsenic is fairly constant over time (Concha 2001). Therefore, interindividual variation cannot be explained by the fact that in most studies there is only one urine sample per person.

The variation among individuals seems to be more marked in the amount of urinary MMA than in the amount of DMA. The percentage of MMA might vary 30-fold among individuals in a particular population, whereas the variation in the percentage of DMA is generally no more than 2-fold (Figure 4-2). The observed differences in the urinary excretion of arsenic metabolites might indicate a genetic polymorphism in the regulation of enzymes responsible for arsenic methylation. Genetic polymorphism has been demonstrated for other human methyltransferases (Weinshilboum et al. 1999).

Effect of Dose on Arsenic Methylation

Experimental animal studies have shown that several factors, including dose, influence the methylation of inorganic arsenic (NRC 1999). Recent studies are consistent with the effect of dose on arsenic methylation reported in the previous report (NRC 1999). In Region Lagunera, Mexico, children with an average of 70% of urinary arsenic as DMA showed a significant decrease in the percentage of DMA and a concurrent increase in the percentage of inorganic arsenic and MMA in urine with increasing exposure to arsenic (total urinary arsenic at 100-1,100 µg/L)(Del Razo et al. 1999). In a few children with some of the highest exposures, only 40-50% of urinary arsenic was DMA, MMA constituted 20%, and inorganic arsenic 25% of total urinary arsenic. It is probable that the decrease in methylation with increasing dose is due to inhibition of methyltransferase. As discussed in Chapter 3, recent experimental studies in vitro confirm previous findings that methylation of arsenic, especially the second methylation step in which MMA is converted to DMA, is inhibited by excess amounts of arsenic (Styblo et al. 1999, 2000; De Kimpe et al. 1999).

Effect of Other Chemicals on Arsenic Methylation

Arsenic methylation might also be influenced by simultaneous exposure to other exposures or chelating agents. Experimental studies indicate that the methylation of inorganic arsenic to MMA is inhibited by vanadium, iron, selenite, and cadmium; specifically, selenite, mercury, lead, and chromium inhibit the second methylation step (De Kimpe et al. 1999). Administration of chelating agents, such as DMPS, has been found to change the urinary arsenic metabolite pattern dramatically (Aposhian et al. 2000a,b; Le et al. 2000). In particular, urinary excretion of MMA increased. Of total arsenic in urine, the percentage of MMA increased by about 10-fold and the percentage of DMA decreased following treatment with DMPS.

In vitro studies using liver cytosol from the Flemish rabbit demonstrated that British Anti-Lewisite (BAL), dimercaptosuccinic acid (DMSA), and ethylenediaminetetraacetic acid (EDTA), all at concentrations between 0.15 and 15 mM, inhibit the methylation of both arsenite and MMA (De Kimpe et al. 1999). DMPS mainly inhibited the methylation of MMA to DMA. Citrate had a limited stimulatory effect on both steps in the arsenic methylation.

Metabolic Variability, Tissue Retention, and Health Effects

If the reactive metabolite MMA[III] is formed and distributed to the tissues following exposure to inorganic arsenic, it seems likely that people who excrete a relatively higher proportion of MMA in the urine might have a higher retention of the absorbed arsenic than those who excrete lesser amounts of MMA. Indeed, most animals for which the main product of inorganic arsenic methylation is DMA show a faster overall excretion of arsenic than do humans (Vahter 1999a). In Andean people, who have only a small percentage of urinary arsenic as MMA, the ratio of arsenic in blood to arsenic in urine (adjusted for variation in total urinary metabolites of inorganic arsenic) was very low (about 0.03) (Concha et al. 1998). In contrast, in a study that measured total urinary arsenic in people from California and Nevada, that ratio was twice as high (about 0.06) (Valentine et al. 1979). Although the urinary metabolites of arsenic were not reported in the latter study, the urine can be assumed to contain 10-20% MMA, because recent studies of people exposed to arsenic in drinking water in Nevada showed 20% inorganic arsenic, 22% MMA, and 58% DMA in the urine (Warner et al. 1994). Because arsenic from seafood is readily excreted, seafood arsenic influences urine concentrations more than blood concentrations. Therefore, measuring total urinary arsenic, rather than metabolites of inorganic arsenic in the urine, might underestimate the true ratio of blood-to-urine arsenic from inorganic arsenic. Thus, the ratio in the study by Valentine et al. (1979) might have been even higher.

A lower retention of arsenic in people with low MMA excretion is also suggested by data from a number of experimental studies on human subjects receiving specified doses of inorganic arsenic (Vahter 2001). The percentages of urinary arsenic as MMA and inorganic arsenic showed similar negative associations with the overall arsenic excretion, while the percentage of urinary arsenic as DMA showed a positive association. Thus, urinary excretion of all arsenic metabolites (percentage of dose) decreased with increasing percentage of both inorganic arsenic and MMA in urine but increased with increasing percentage of DMA.

A number of recent studies indicate an association between the prevalence of arsenic-related toxic effects and the pattern of arsenic metabolites in urine. In particular, there are indications that people with arsenic-related toxic effects have a higher percentage of urinary arsenic as MMA compared with people without such effects. Mexican people with signs of dermal toxicity due to exposure to arsenic in drinking water had a higher percentage of urinary arsenic MMA and a lower percentage of DMA than those without visible dermal toxicity (Del Razo et al. 1997). The difference was about 5% MMA. In the area of southwestern Taiwan where blackfoot-disease is endemic, skin cancer patients were found to have a higher percentage of urinary arsenic as MMA, 31% on average compared with 23% among the controls (Hsueh et al. 1997). Skin-cancer patients were also found to have lower serum β-carotene concentrations than controls. However, the association between percentage of urinary arsenic as MMA and the incidence of skin cancer remained significant after adjustment for cumulative arsenic exposure and serum β-carotene concentrations. In a later study from the same area of Taiwan, a significantly higher percentage of urinary arsenic as inorganic arsenic (about 2%) and MMA (about 2%) and a lower percentage as DMA were found in people with arsenic-related skin lesions than in matched controls (Yu et al. 2000). The number of structural chromosomal aberrations in peripheral lymphocytes of Finnish people exposed to arsenic via drinking water (average concentration of 410 μg/L) was also found to be associated with increasing percentage of urinary arsenic as MMA and decreasing percentage as DMA (Mäki-Paakkanen et al. 1998). Further studies are needed to clarify the roles of MMA and DMA in arsenic toxicity.

SOURCES OF UNCERTAINTY

Dose Metrics and Model Uncertainty

The association between arsenic exposure and an adverse health effect can be calculated by comparing the response to a range of different dose metrics, including cumulative arsenic dose, average daily intake over a lifetime, and peak arsenic exposure, and using a wide range of different statistical models. The dose metric and model used can affect the calculation of risk estimates, and depending on how accurately exposure is measured by the dose metric, and how closely the metric is related to the end point of concern, more or less uncertainty is introduced to a risk assessment. The end point of concern for arsenic toxicity is cancer (see Chapter 2). In its risk impact assessment, EPA (2000a) stated,

> In certain circumstances, the increased daily dose in children can be effectively considered for non-carcinogenic effects because toxicity is evaluated in terms of exposure that can range from relatively short-term to long-term exposure. However, carcinogenic effects (i.e., bladder cancer) are evaluated based on a lifetime of exposure, which takes into consideration the elevated dose that occurs in children. Because the health effects measured in this benefits assessment [EPA's Risk Impact Assessment] are bladder and lung cancer, a sensitivity analysis to consider higher doses of arsenic during childhood was not necessary.

The subcommittee agrees with that statement but emphasizes that its validity also depends on the lifetime cancer risk focus that uses the lifetime average daily dose as the dose metric. The subcommittee further emphasizes that cancer represents the most sensitive health end point and that the dose-response data are based on a population with lifetime exposure to arsenic. Thus, the dose-response model reflects lifetime exposure (from childhood to adulthood), as expected in any model for lifetime cancer risk. Clearly, uncertainty exists about whether this model is appropriate, as discussed more in Chapter 5. As noted in Chapter 3, at a mechanistic level, little is known about the impact of prolonged exposure at relatively low doses compared with shorter exposures at high doses, and this lack of knowledge arises as an important source of uncertainty in assessing the risks of ingesting arsenic. The lack of agreement between health effects observed in the epidemiological studies described in Chapter 2 and those observed in animals described in Chapter 3 raises additional uncertainties about the appropriate dose-response model and dose metric for humans when using animal data. If the peak concentration represents the appropriate dose metric of concern, then use of a lifetime cancer risk model would not be valid. Assuming that environmental exposures are essentially constant, peak concentrations likely occur when people are infants because of the higher water intake per unit of body weight. However, if cumulative dose is important, or if arsenic acts as a late-stage carcinogen, information about the timing and amount of exposure is needed.

At this time, the mode of action of arsenic and the dose metrics that are best correlated with the carcinogenic effects of exposure to arsenic in drinking water are not known. The dose metric affects the interpretation of an epidemiological study. It is possible that the lack of an association between arsenic and cancer could result from using an inappropriate dose metric. For example, if the duration and magnitude of exposure are both important for arsenic-induced cancer, then cumulative exposure might be associated with cancer, and peak exposure might not be associated with cancer. The issue of dose metrics, therefore, complicates the epidemiological studies and must be considered when interpreting and comparing different studies. Because the most appropriate dose metric for arsenic-induced cancer is still not known, the choice of metric adds uncertainty to arsenic risk assessments.

In addition, a wide range of different models can be used to fit the arsenic carcinogenicity data currently available, and no clear biological basis exists for distinguishing among them. The implications of model uncertainty are explored quantitatively in Chapter 5.

Exposure to Arsenic in Food and Effect of Diet and Nutrition

In addition to adjusting for an assumed higher drinking-water rate among Taiwanese people compared with the U.S. population in its most recent risk assessment, EPA (2001) also adjusted for differences in dietary arsenic intake. This type of adjustment is appropriate since arsenic in the diet represents an important source of exposure, and one that must be considered in the context of understanding the dose-response data observed in the ecological epidemiological studies, because it contributes to the total dose. Since the EPA and the subcommittee have focused on use of the Taiwanese epidemiological data as the basis for developing a quantitative dose-response model, the estimation of the proportion of the total Taiwanese arsenic intake that resulted from food emerges as a source of uncertainty.

The previous NRC report on arsenic (NRC 1999) had an extensive discussion about arsenic in food. It was concluded that the highest concentrations of arsenic are found in products from the marine environment. In fish and shrimp, the major arsenical is arsenobetaine. In most marine algal products, arsenosugars are the principal arsenic species, although up to 50% might be the more toxic arsenate. Arsenosugars are metabolized by humans mainly to DMA. Although the subcommittee was not asked to evaluate the exposure to arsenic, it notes that the recent Food and Drug Administration Total Dietary Study for 1991-1997 (Tao and Bolger 1999) indicates that a major source of arsenic in the U.S. diet is of marine origin. In that study, total inorganic and organic arsenic in food was detected in 63 of the 261-264 (24%) foods and mixed dishes analyzed at a detection limit of 0.03 ppm. The highest concentration was found in seafood, followed by rice and rice cereal, mushrooms, and poultry.

Based on the U.S. Department of Agriculture's 1987-1988 Nationwide Food Consumption Survey, the estimated daily total arsenic average intakes were 2 µg/day for infants, 23 µg/day for toddlers, 20 µg/day for 6-year-old children, 42 µg/day for adults (40-45 years of age), and 82 µg/day for individuals 60-65 years of age. Of the estimated total arsenic intakes for infants, 42% originated from seafood and 31% from rice and rice cereals. Of the estimated total arsenic intakes, seafood contributed 76% to 90% for children (2-10 years old) and 89% to 96% for adults (25-30 years old), whereas rice and rice cereals contributed 4% to 8% for children and 1% to 4% for adults (25-30 years old).

In a recent market basket survey, 40 commodities anticipated to provide at least 90% of dietary inorganic arsenic intake were identified (Schoof et al.

1999). Total arsenic was analyzed using NaOH digestion and inductively coupled plasma-mass spectrometry. Inorganic arsenic was analyzed using HCl digestion and hydride atomic absorption spectroscopy. Consistent with earlier studies, total arsenic concentrations (all concentrations reported as elemental arsenic per tissue wet weight) were highest in seafood (ranging from 160 ng/g in freshwater fish to 2,360 ng/g in saltwater fish). In contrast, average inorganic arsenic in seafood ranged from less than 1 to 2 ng/g. The highest inorganic arsenic values were found in raw rice (74 ng/g), followed by flour (11 ng/g), grape juice (9 ng/g), and cooked spinach (6 ng/g). Thus, grains and produce are expected to be significant contributors to dietary inorganic arsenic intake.

Based on these data and previous studies on concentrations of arsenic metabolites in urine (NRC 1999), the average dietary intake of inorganic arsenic in the United States is likely to be on the order of 10 µg/day, as estimated by EPA. Also, the dietary intake of inorganic arsenic in the area of southwest Taiwan where arsenic is endemic is possibly higher than that in the United States, because although the water with increased arsenic concentrations usually is no longer ingested, it is still used for agriculture and pisciculture, as well as washing dishes, cleaning, watering plants, and, occasionally, drinking in dry seasons (Hsueh et al. 1997; Yu et al. 2000). The few available data on arsenic in rice and yams from Taiwan were reported in the NRC 1999 report.

However, there is little evidence on the levels and species of arsenic consumed in foods by different individuals and populations, and the role of arsenic in food remains somewhat uncertain. Furthermore, while the new data should be used in the context of estimating total arsenic exposure and risk to the U.S. population, they are not particularly helpful for the subcommittee's efforts to characterize the amounts of arsenic in the food of the historically exposed Taiwanese population. Such data are needed to appropriately characterize the dose-response function for total ingested arsenic. The bioavailability to humans of arsenic present in raw foods and the bioavailability of arsenic incorporated into foods, such as rice from cooking water, has not been well characterized.

In its risk assessment, EPA assumed that food intake of arsenic was higher in Taiwan than in the United States. However, there is little evidence to support the numbers used, and the subcommittee explores this important source of uncertainty in Chapter 5.

Also related to diet is the hypothesis that dietary and nutritional aspects of the Taiwanese population render it more susceptible to the carcinogenic effects of arsenic than the U.S. population and, therefore, that the results of the Taiwanese studies might not be relevant to the U.S. population. This hypothesis is based in part on the fact that cancer has been consistently associated with aspects of diet; both increased and decreased cancer risks have been associated with different dietary parameters (Doll and Peto 1981; NRC 1982; Steinmetz and Potter 1991a,b; World Cancer Research Fund Panel 1997). Particularly relevant to the arsenic and cancer data are the epithelial cancers, in which the association between diet and cancer is most obvious (World Cancer Research Fund Panel 1997).

Before discussing the data specific to arsenic, it is important to consider the potential influence of diet in a general sense. Deciding on the strength of the association between an exposure and an outcome is heavily influenced by the precision with which the exposure can be measured. Diet and eating habits are complex and difficult to measure accurately (Willett 1990). Typically, when errors in exposures occur, the derived estimates of risk are biased toward the null; that is, the association, as estimated, is weaker than it would be with perfect measurement. The degree of bias in the effect of food on cancer incidence remains a matter of considerable debate, but it is possible that diet could affect the epidemiological studies of arsenic. There are three general ways by which diet and nutrition could affect the association between arsenic and diseases (including cancers), particularly in the Taiwanese population studies: (1) confounding by diet; (2) potentiation of susceptibility as a result of poor-quality diet; and (3) the difficulty with generalizing the Taiwanese findings to other populations because of differences in diet.

Confounding occurs when there is a condition that might be a factor in producing the same response as the agent of interest. Although some uncertainties remain, the subcommittee believes that based on the existing weight of the evidence, it is unlikely that confounding by diet is responsible for the association between arsenic exposure and cancer. The association between exposure to arsenic and cancer has been seen in several independent investigations conducted in different populations with varying diets, using a variety of study designs. The magnitudes of the observed relative risks for cancer were in general so high that they could not be accounted for by known or hypothesized dietary factors. For example, in the analysis by Smith et al. (1992) of the southwestern Taiwanese data, the estimated mortality risk ratio for bladder cancer in females relative to the general Taiwanese population increased in a dose dependent manner from 11.9 (water arsenic concentration less than 300 μg/L), to 25.1 (300 to 600 μg/L) to ultimately 65.4 (greater than 600 μg/L). In the study by Chiou et al. (2001) in northeastern Taiwan that used a prospective study design with individual exposure data and internal controls, there was a 15-fold increase in risk for bladder cancer in the subjects (male and female combined) with the highest levels of exposure.

A more general consideration is that although diet remains unmeasured in most of the epidemiological studies of arsenic and cancer, when it has been measured in other settings, it is almost never a confounder of relationships between specific and well-measured exposures. For example, diet has not been identified as a significant confounder of the relationship between benzene, asbestos, or cigarette smoking and relevant specific cancer outcomes. Nonetheless, in the absence of clear evidence of a biological mechanism for arsenic-induced carcinogenesis (see Chapter 3), the possibility of a relatively minor degree of unmeasured confounding remains, even if diet is an unlikely candidate.

In addition to confounding, interaction or modification of effects might also occur when the association between exposure and outcome is much stronger among those with, versus without, a different characteristic. For example, smokers might have a 4-fold increase in risk for a particular disease, and those exposed to radiation might have a doubling of risk. If smokers who are also exposed to radiation have a 20-fold increase in disease, there is evidence of interaction, which suggests biological synergism of one exposure with another.

In the case of arsenic, interaction or modification of effects by poor nutrition could potentially increase susceptibility. In the southwestern Taiwanese population, the staple foods used to be sweet potatoes and rice, with a low intake of vegetables and fruit and their bioactive constituents that can protect against cancer. The impact of nutrition on arsenic susceptibility in the study populations was discussed in detail in the 1999 report, and more research has been conducted since that time. In a recent study from West Bengal, India (see Dermal Effects in Chapter 2), it was demonstrated that among people exposed to high concentrations of arsenic in drinking water, those who were below 80% of standard body weight had a 1.6-fold increase in the prevalence of keratosis (Mazumder et al. 1998). In a case-control study of 241 blackfoot-disease (BFD) cases in southwestern Taiwan, artesian-well-water consumption, arsenic poisoning, familial history of BFD, and undernourishment were all significantly associated with the development of BFD (Chen et al. 1988). Undernourishment was mainly characterized by higher intake of dried sweet-potato chips (the major food for rural residents in southwestern Taiwan before 1960) and lower intake of rice and vegetables. According to the study by Hsueh et al. (1995), the prevalence of arsenic-induced skin cancer increased with increasing consumption of dried sweet potatoes as staple food. It also increased with chronic liver disease (chronic carriers of hepatitis B antigen with liver dysfunction). Those data support earlier findings that malnutrition might increase susceptibility to arsenic (NRC 1999).

Recent studies argue against the assertion that poor nutrition has a major impact on the toxicity of arsenic. The prevalence of arsenic-induced skin lesions in northern Chile was reported to be similar to that for areas of Taiwan and West Bengal. However, although the areas had similar arsenic concentrations in the drinking water, the study population in Chile had good nutrition and those in the other areas did not (Smith et al. 2000). In Chile, 4 of 11 men (36%) and 2 of 23 children (9%) had skin effects in the form of pigmentation changes and hyperkeratosis. These data were compared with prevalences of skin effects of 11% among men (30-59 years) and 5% among children less than 20 years of age in West Bengal (Mazumder et al. 1998). They were also compared with an overall prevalence of 18% hyperpigmentation and 7% hyperkeratosis in Taiwan, although age distribution was not presented (Tseng et al. 1968; Tseng 1989). However, it should be noted that the study in Chile was limited to 11 exposed families and 8 control families.

Taken together, those data indicate that nutritional status might influence the risk for health effects following arsenic exposure. Thus, the risk estimates based on the population in southwestern Taiwan might have been influenced by the poor nutrition of that population at the time of the increased arsenic exposures. However, the recent studies from northeastern Taiwan (Chiou et al. 2001) and Chile (Ferreccio et al. 2000) that can also be used for quantitative risk characterization (see Chapter 5) involve populations with much better nutrition.

Some research has investigated whether the status of specific nutrients is associated with an increased susceptibility to arsenic-induced cancer. In the area of southwestern Taiwan where blackfoot-disease is endemic, skin-cancer patients were found to have significantly lower serum β-carotene concentrations compared with controls (Hsueh et al. 1997). The odds ratio for skin cancer decreased from 1.0 at a serum β-carotene concentration of 0.14 mg/L or less to 0.43 (95% confidence interval (CI) = 0.06-2.85) at 0.15-0.18 mg/L and to 0.01 (95% CI = 0.00-0.37) at greater than 0.18 mg/L. A synergistic interaction was seen between duration of consumption of artesian well water with high arsenic concentrations and low serum β-carotene concentrations in the development of ischemic heart disease (Hsueh et al. 1998).

Selenium status has also been suggested as an influence on the toxicity of arsenic (see NRC 1999), and data in laboratory animals show that relatively high doses of selenium can affect arsenic metabolism. Studies on mice fed diets with various relatively high doses of selenium indicate that the animals on the selenium-deficient diet had a slower elimination of single oral doses of arsenite, arsenate, or DMA compared with selenium-sufficient mice (Kenyon et al. 1997). Mice fed a diet with excess selenium showed a decreased methylation of inorganic arsenic. Similarly, in vitro studies using primary rat hepatocytes showed that concurrent exposure to selenite (0.1 to 6 μM) inhibited the methylation of arsenite (Styblo and Thomas 2001). The second methylation step, MMA to DMA, seemed to be more sensitive to the inhibitory effect of selenium as the ratio of DMA to MMA decreased significantly with selenium treatment.

Given that a possible mechanism of action of arsenic involves the influence of arsenic on DNA repair, low folate intake could cause increased susceptibility to arsenic carcinogenesis (EPA 1997). Deficiency in folate and vitamin B12 might lead to decreased levels of S-adenosylmethyltransferase (SAM), increased levels of serum homocysteine, and possibly hypomethylation (Newman 1999). Thus, deficiency in those vitamins might result in decreased methylation of arsenic. Folic acid was found to protect SWV/Fnn mouse embryo fibroblasts against cytotoxicity of arsenite and DMA (Ruan et al. 2000), but the folic acid had less effect on the arsenic-related growth inhibition. However, even if deficiencies of folate and other micronutrients increase susceptibility to arsenic, intakes of many micronutrients are often low in developed countries, particularly among the poor; therefore, that increased susceptibility might be relevant to a segment of the U.S. population.

The final question is whether the Taiwanese findings can be generalized to the United States even though the populations have different patterns of dietary intake. There are two issues here. First, the pathobiology of arsenic -related deleterious outcomes, including carcinogenesis, appears to be consistent across human populations, with little evidence of selection for resistance to some consequences even with millenia of exposure (Smith et al. 2000). Second, although human diets vary substantially, patterns of cancer risk and protection against cancer are also relatively consistent across populations (World Cancer Research Fund Panel 1997). As suggested above, there might be variation across countries in the proportion of the population exposed to carcinogens or protected against those carcinogens by host biology or protective dietary intake. The pathobiology itself is relatively uniform.

Other Uncertainties

Although the discussion on latency in Chapter 2 mentioned a number of uncertainties, it is worthwhile to reiterate that the subcommittee's understanding of the timing of transitions on the path from exposure to disease for arsenic remains highly uncertain, even as more is learned about the underlying mechanisms of disease. In addition, the variability in latency of different types of arsenic-related diseases has not been explored. These might arise from a difference in the exposure level, duration of exposure, age of exposure, susceptibility (that might lead some people to respond more quickly than others), or other unknown risk factors.

In assessing the risks from arsenic, it is nearly impossible to entirely rule out the possibility that genetics, lifestyle differences (e.g., smoking, food preferences, cooking habits), and exposure to other environmental factors might play a role in explaining variability in the risks. A few investigations have suggested that there might be an interaction between arsenic exposure and smoking in cancer causation. A case-control study by Bates et al. (1995) examined the relationship between arsenic exposure in drinking water and bladder cancer and found a positive association between arsenic and bladder cancer only among smokers. The findings of two studies (Ferreccio et al. 2000; Tsuda et al. 1995) suggest that there might be a synergistic effect between the arsenic ingestion and smoking on the risk of lung cancer. A synergistic interaction for lung cancer between smoking and inhaled arsenic has been noted in occupational cohorts (Hertz-Picciotto et al. 1992). Further confirmation and characterization of an interaction between arsenic ingestion and smoking in the causation of cancer are necessary before this potential effect can be accounted for in a quantitative risk assessment.

Nevertheless, it is important to emphasize that there is little reason to suspect that smoking is a significant confounder of the association between arsenic ingestion and lung or bladder cancer. In the case-control study of lung cancer in Chile by Ferreccio et al. (2000) described in Chapter 2, high odds ratios for lung cancer with increasing arsenic exposure were observed in models that adjusted for smoking history. In the cohort study of incident bladder cancer in northeastern Taiwan reported by Chiou et al (2001), the observed association with arsenic exposure also persisted after adjusting for smoking history. As noted by NRC (1999), an ecological study by Smith et al. (1998) reported standardized mortality ratios (SMRs) for lung and bladder cancer in a region of northern Chile (Region II) for past exposures to high arsenic concentrations in drinking water. The SMR for chronic obstructive pulmonary disease (COPD) in females in that region was 0.6 (95% CI = 0.4-0.7). Since COPD is overwhelmingly due to smoking, it is apparent that the region contained few female smokers relative to the rest of Chile. Yet the female lung cancer SMR in that region was 3.1 (95% CI = 2.7-3.7), and the female bladder cancer SMR was 8.2 (6.3-10.5). As noted by the authors, there was a large risk of arsenic associated lung and bladder cancer that was not subject to confounding by smoking. A similar finding emerged for males. The high SMRs for lung cancer and bladder mortality in the region of southwest Taiwan where arsenic is endemic were unlikely to be subject to significant confounding by smoking or other factors (see description of Tsai et al. (1999) in Chapter 2). Neither males nor females in the area where arsenic is endemic had elevated SMRs for emphysema, a cause of death almost entirely attributable to smoking.

VALUE-OF-INFORMATION APPROACH

Future research might reveal that the manifestations of arsenic toxicity are influenced by gene-environment interactions or that some populations might be at relatively increased risk of developing cancer from exposure to arsenic based on their genetics or their behaviors (e.g., smokers). The implications of this type of variability will raise important policy considerations for risk managers, and research to support the efforts of decision-makers to deal with these issues should be initiated.

In a 1996 report, the National Research Council discussed the importance of using an analytic-deliberative process in the context of managing risks (NRC 1996). As noted in the report of the Presidential/Congressional Commission on Risk Assessment and Risk Management (1997), better risk-management decisions will be made when the process has the capacity for iterations "if new information is developed that changes the need for or nature of risk management." A value-of-information approach can be used to decide if it is more appropriate to collect additional information or make a decision based on a risk assessment with specified uncertainties. Given the existing uncertainties that remain for arsenic, EPA should explore whether implementing a value-of-information approach to prioritizing research efforts would be helpful in supporting its process for regulating arsenic. As discussed by NRC (1996),

> Value-of-information methods address whether potential reductions in uncertainty would make a difference in the decision; they suggest priorities among reducible uncertainties on the basis of how much difference the expected reduction might make (p. 110).

Although the decision to implement a value-of-information approach reflects a policy choice, it might help to focus research efforts on those key uncertainties that would have the largest impact on our understanding of the magnitude of arsenic risks.

SUMMARY AND CONCLUSIONS

The subcommittee commends EPA for considering variability and uncertainty in its risk assessment for arsenic and hopes that EPA will continue to refine and update its assessment. New data developed since the 1999 report provide more insight into how sources of variability and uncertainty can be addressed in a risk assessment for arsenic. For example, variability in drinking-water rates per unit of body weight can now be quantitatively analyzed. The subcommittee explores the implications of using that type of information on variability in risk assessment in Chapter 5. However, there are still a number of key sources of uncertainty that have not been addressed. They include the following:

- The mechanisms by which arsenic causes cancer are not well understood. It is unclear what the shape of its dose-response curve is at low doses and whether the magnitude of the dose or the duration of exposure is more important in cancer risk.
- There is a lack of concordance in the results of animal and human studies.
- Geographical and cultural differences in arsenic exposures make it difficult to relate consumption habits in Taiwan with those in the United States. To make better comparisons and account for differences, more information is needed on arsenic content in foods, water-ingestion rates, food-preparation practices, and nutritional status of the populations. Dietary and nutritional variability between population groups might result in a different proportion of the populations being at risk for arsenic-induced health effects, but the risk will be increased in all those who share the same pattern of exposure and susceptibility.
- It is unclear whether infants and young children might be more susceptible to arsenic-induced health effects, particularly those for noncancer end points where less-than-lifetime exposures are important and children's greater water consumption per unit of body weight might put them at relatively greater risk.
- The metabolism and extent of methylation of arsenic is not fully understood. For example,
— There is considerable interindividual variation in arsenic metabolism, particularly in the production of MMA and DMA. Some data indicate that methylation is influenced more by genetic factors than by environmental factors. Recent studies, however, confirm the previous findings that methylation of arsenic, especially the second methylation step in which MMA is

converted to DMA, is inhibited at higher doses of arsenic. Data also indicate less-efficient methylation of arsenic to DMA in children compared with adults, but those data are not conclusive.

— Experimental studies indicate that exposure to other pollutants (e.g., vanadium, selenium, cadmium, mercury, and lead) might inhibit the second arsenic methylation step. Administration of chelating agents, such as DMPS, has been found to decrease the urinary excretion of DMA and increase that of MMA.

— Arsenic metabolites vary considerably in their toxicity; thus, variation in the metabolism of arsenic is likely to be associated with variations in susceptibility to arsenic. Animal and human data indicate that efficient methylation of inorganic arsenic to DMA^V results in faster overall excretion of arsenic. Also, there is increasing evidence that individuals with increased MMA production and retention (mainly MMA^{III}) retain more arsenic and are more prone to at least some toxic effects of arsenic.

RECOMMENDATIONS

• EPA should explore research opportunities to reduce the key uncertainties identified and should support research efforts that might resolve important uncertainties. EPA should also explore using an iterative value-of-information approach to prioritize future research efforts targeted at resolving uncertainties in the arsenic risk assessment. When better data and information are obtained, they should be factored in the risk assessment.

• In considering how variability in the amount of drinking water consumed would affect quantitative risk assessment for arsenic, EPA should adjust for variability in consumption in the population that was the source of the data (e.g., the southwestern Taiwanese), as well as in the general U.S. population.

• Key sources of uncertainty that may be subject to substantial reduction include the following:

— Better understanding of the differences in arsenic exposure between U.S. and Taiwanese populations. This information could be obtained through surveys of the Taiwan drinking-water consumption, water usage for cooking, and body weights. However, it is unclear whether data obtained at the present time could be applied retrospectively to a study population whose key exposure occurred decades in the past.

— Improved characterization of the form and bioavailability of the arsenic present in the raw foods and of the arsenic incorporated into food from drinking water during cooking and food preparation. Such characterization will enhance knowledge of valid means to account for this arsenic in quantitative risk assessment, which is particularly relevant to EPA's exposure assessment efforts.

— Clarification of the mechanisms of action of arsenic and development of an appropriate animal model for studying the effects of arsenic. Such clarification and development would help to reduce the uncertainty about dose metrics and the relevance of animal data for human risk assessments.

• Further characterization of the long-term drinking-water consumption patterns in the general population of the United States could narrow the current characterization of variability in drinking-water rates. However, although it might reduce the overall variability in the estimated risks, it is unlikely to dramatically change the risk estimates.

• More information is needed on the variability in the metabolism of arsenic among individuals and the effect of that variability on an individual's susceptibility to cancer and systemic toxicity. In particular, the impact of variability in MMA[III] formation and distribution in human tissues is needed. Factors influencing the variability, genetic as well as environmental, should be studied.

• More data are needed to better understand the susceptibility of children to arsenic-induced toxicity, particularly for noncancer effects, and their arsenic-methylation efficiency.

• The influence of nutritional and dietary factors on the risk for arsenic-induced health effects should be subject to further research.

REFERENCES

Aposhian, H.V., A. Arroyo, M.E. Cebrian, L.M. Del Razo, K.M. Hurlbut, R.C. Dart, D. Gonzalez-Ramirez, H. Kreppel, H. Speisky, A. Smith, M.E. Gonsebatt, P. Ostrosky-Wegman, and M.M. Aposhian. 1997. DMPS-arsenic challenge test. I. Increased urinary excretion of monomethylarsonic acid in humans given dimercaptopropane sulfonate. J. Pharmacol. Exp. Ther. 282(1):192-200.

Aposhian, H.V., B. Zheng, M.M. Aposhian, X.C. Le, M.E. Cebrian, W. Cullen, R.A. Zakharyan, M. Ma, R.C. Dart, Z. Cheng, P. Andrewes, L. Yip, G.F. O'Malley, R.M. Maiorino, W. Van Voorhies, S.M. Healy, and A. Titcomb. 2000a. DMPS – Arsenic challenge test. II. Modulation of arsenic species, including monomethylarsonous acid (MMA[III]), excreted in human urine. Toxicol. Appl. Pharmacol. 165(1):74-83.

Aposhian, H.V., E.S. Gurzau, X.C. Le, A. Gurzau, S.M. Healy, X. Lu, M. Ma, L. Yip, R.A. Zakharyan, R.M. Maiorino, R.C. Dart, M.G. Tircus, D. Gonzalez-Ramirez,

D.L. Morgan, D. Avram, and M.M. Aposhian. 2000b. Occurrence of monomethylarsonous acid in urine of humans exposed to inorganic arsenic. Chem. Res. Toxicol. 13(8):693-697.

Bates, M.N., A.H. Smith, and K.P. Cantor. 1995. Case-control study of bladder cancer and arsenic in drinking water. Am. J. Epidemiol. 141(6):523-530.

Calderon, R.L., E. Hudgens, X.C. Le, D. Schreinemachers, and D.J. Thomas. 1999. Excretion of arsenic in urine as a function of exposure to arsenic in drinking water. Environ. Health Perspect. 107(8):663-667.

Chen, C.J., Y.C. Chuang, T.M. Lin, and H.Y. Wu. 1985. Malignant neoplasms among residents of a blackfoot disease-endemic area in Taiwan: high-arsenic artesian well water and cancers. Cancer Res. 45(11 Pt 2):5895-5899.

Chen, C.J., Y.C. Chuang, S.L. You, T.M. Lin, and H.Y. Wu. 1986. A retrospective study on malignant neoplasms of bladder, lung and liver in blackfoot disease endemic area of Taiwan. Br. J. Cancer 53(3):399-405.

Chen, C.J., M.M. Wu, S.S. Lee, J.D. Wang, S.H. Cheng, and H.Y. Wu. 1988. Atherogenicity and carcinogenicity of high-arsenic artesian well water. Multiple risk factors and related malignant neoplasms of blackfoot disease. Arteriosclerosis 8(5):452-460.

Chen, C.J., C.W. Chen, M.M. Wu, and T.L. Kuo. 1992. Cancer potential in liver, lung, bladder and kidney due to ingested inorganic arsenic in drinking water. Br. J. Cancer 66(5):888-892.

Chiou, H.Y., Y.M. Hsueh, L.L. Hsieh, L.I. Hsu, Y.H. Hsu, F.I. Hsieh, M.L. Wei, H.C. Chen, H.T. Yang, L.C. Leu, T.H. Chu, C. Chen-Wu, M.H. Yang, and C.J. Chen. 1997. Arsenic methylation capacity, body retention, and null genotypes of glutathione S-transferase M1 and T1 among current arsenic-exposed residents in Taiwan. Mutat. Res. 386(3):197-207.

Chiou, H.Y., S.T. Chiou, Y.H. Hsu, Y.L. Chou, C.H. Tseng, M.L. Wei, and C.J. Chen. 2001. Incidence of transitional cell carcinoma and arsenic in drinking water: a follow-up study of 8,102 residents in an arseniasis-endemic area in northeastern Taiwan. Am. J. Epidemiol. 153(5):411-418.

Concha, G., B. Nermell, and M. Vahter. 1998. Metabolism of inorganic arsenic in children with chronic high arsenic exposure in northern Argentina. Environ. Health Perspect. 106(6):355-359.

Concha, G.Q. 2001. Metabolism of Inorganic Arsenic and Biomarkers of Exposure. Doctoral Thesis. Institute of Environmental Medicine, Karolinska Institutet, Stockholm.

De Kimpe, J., R. Cornelis, and R. Vanholder. 1999. In vitro methylation of arsenite by rabbit liver cytosol: effect of metal ions, metal chelating agents, methyltransferase inhibitors and uremic toxins. Drug. Chem. Toxicol. 22(4):613-628.

Del Razo, L.M., G.G. García-Vargas, H. Vargas, A. Albores, M.E. Gonsebatt, R. Montero, P. Ostrosky-Wegman, M. Kelsh, and M.E. Cebrián. 1997. Altered profile of urinary arsenic metabolites in adults with chronic arsenicism. A pilot study. Arch. Toxicol. 71(4):211-217.

Del Razo, L.M., G.G. García-Vargas, M.C. Hernández, Gómez-Muñoz, and M.E. Cebrián. 1999. Profile of urinary arsenic metabolites in children chronically exposed to inorganic arsenic in Mexico. Pp. 281-287 in Arsenic Exposure and Health Effects, W.R. Chappell, C.O. Abernathy, and R.L. Calderon, eds. Oxford: Elsevier.

Doll, R., and R. Peto. 1981. The causes of cancer: quantitative estimates of avoidable risk of cancer in the United States today. J. Natl. Cancer Inst. 66(6):1191-1308.

EPA (U.S. Environmental Protection Agency). 1988. Special Report on Ingested Inorganic Arsenic: Skin Cancer; Nutritional Essentiality. EPA/625/3-87/013. Risk Assessment Forum, U.S. Environmental Protection Agency, Washington, DC. July 1988.

EPA (U.S. Environmental Protection Agency). 1997. Report of the Expert Panel on Arsenic Carcinogenicity: Review and Workshop. Prepared by Eastern Research Group, Lexington, MA, for the U.S. Environmental Protection Agency, National Center for Environmental Assessment, Washington, DC.

EPA (U.S. Environmental Protection Agency). 2000a. 40 CFR Parts 141 and 142. National Primary Drinking Water Regulations. Arsenic and Clarifications to Compliance and New Source Contaminants Monitoring. Notice of proposed rulemaking. Fed. Regist. 65(121):38887-38983.

EPA (U.S. Environmental Protection Agency). 2000b. Estimated Per Capita Water Ingestion in the United States: Based on Data Collected by the United States Department of Agriculture's (USDA) 1994-1996 Continuing Survey of Food Intakes by Individuals. EPA-822-00-008. Office of Water, Office of Standards and Technology, U.S. Environmental Protection Agency.

EPA (U.S. Environmental Protection Agency). 2001. 40 CFR Parts 9, 141 and 142. National Primary Drinking Water Regulations. Arsenic and Clarifications to Compliance and New Source Contaminants Monitoring. Final Rule. Fed. Regist. 66(14): 6975-7066.

Ershow, A.G., L.M. Brown, and K.P. Cantor. 1991. Intake of tapwater and total water by pregnant and lactating women. Am. J. Public Health 81(3):328-334.

Ferreccio, C., C. Gonzalez, V. Milosavljevic, G. Marshall, A.M. Sancha, and A.H. Smith. 2000. Lung cancer and arsenic concentrations in drinking water in Chile. Epidemiology 11(6):673-679.

Hertz-Picciotto, I., A.H. Smith, D. Holtzman, M. Lipsett, and G. Alexeeff. 1992. Synergism between occupational arsenic exposure and smoking in the induction of lung cancer. Epidemiology 3(1):23-31.

Hopenhayn-Rich, C., M.L. Biggs, and A.H. Smith. 1998. Lung and kidney cancer mortality associated with arsenic in drinking water in Córdoba, Argentina. Int. J. Epidemiol. 27(4):561-569.

Hsu, K.H., J.R. Froines, and C.J. Chen. 1997. Studies of arsenic ingestion from drinking-water in northeastern Taiwan: chemical speciation and urinary metabolites. Pp. 190-209 in Arsenic Exposure and Health Effects, C.O. Abernathy, R.L. Calderon, and W.R. Chappell, eds. London: Chapman & Hall.

Hsueh, Y.M., H.Y. Chiou, Y.L. Huang, W.L. Wu, C.C. Huang, M.H. Yang, L.C. Lue, G.S. Chen, and C.J. Chen. 1997. Serum beta-carotene level, arsenic methylation capability, and incidence of skin cancer. Cancer Epidemiol. Biomarkers Prev. 6(8):589-596.

Hsueh, Y.M., Y.L. Huang, C.C. Huang, W.L. Wu, H.M. Chen, M.H. Yang, L.C. Lue, and C.J. Chen. 1998. Urinary levels of inorganic and organic arsenic metabolites among residents in an arseniasis-hyperendemic area in Taiwan. J. Toxicol. Environ. Health A 54(6):431-444.

Hsueh, Y.M., G.S. Cheng, M.M. Wu, H.S. Yu, T.L. Kuo, and C.J. Chen. 1995. Multiple risk factors associated with arsenic-induced skin cancer: effects of chronic liver disease and malnutritional status. Br. J. Cancer 71(1):109-114.

ILSI (International Life Sciences Institute). 1992. Similarities and Differences Between Children and Adults: Implications for Risk Assessment, P.S. Guzelian, C.J. Henry, and S.S. Olin, eds. Washington, DC: ILSI.

IOM (Institute of Medicine). 2001. Dietary Reference Intakes for Vitamin A, Vitamin K, Arsenic, Boron, Chromium, Copper, Iodine, Iron, Manganese, Molybdenum, Nickel, Silicon, Vanadium, and Zinc. Washington, DC: National Academy Press.

Kenyon, E.M., M.F. Hughes, and O.A. Levander. 1997. Influence of dietary selenium on the disposition of arsenate in the female B6C3F1 mouse. J. Toxicol. Environ. Health 51(3):279-299.

Kurttio, P., H. Komulainen, E. Hakala, H. Kahelin, and J. Pekkanen. 1998. Urinary excretion of arsenic species after exposure to arsenic present in drinking water. Arch. Environ. Contam. Toxicol. 34(3):297-305.

Le, X.C., M. Ma, X. Lu, W.R. Cullen, H.V. Aposhian, X. Lu, and B. Zheng. 2000. Determination of monomethylarsonous acid, a key arsenic methylation intermediate, in human urine. Environ. Health Perspect. 108(11):1015-1018.

Mäki-Paakkanen, J., P. Kurttio, A. Paldy, and J. Pekkanen. 1998. Association between the clastogenic effect in peripheral lymphocytes and human exposure to arsenic through drinking water. Environ. Mol. Mutagen. 32(4):301-313.

Mandal, B.K., Y. Ogra, and K.T. Suzuki. 2001. Identification of dimethylarsinous and monomethylarsonous acids in human urine of the arsenic-affected areas in West Bengal, India. Chem. Res. Toxicol. 14(4):371-378.

Mazumder, D.N.G., R. Haque, N. Ghosh, B.K. De, A. Santra, D. Charkraborty, and A.H. Smith. 1998. Arsenic levels in drinking water and the prevalence of skin lesions in West Bengal, India. Int. J. Epidemiol. 27(5):871-877.

Morales, K.H., L. Ryan, T.L. Kuo, M.M. Wu, and C.J. Chen. 2000. Risk of internal cancers from arsenic in drinking water. Environ. Health Perspect. 108(7)655-661.

Mushak, P., and A.F. Crocetti. 1995. Risk and revisionism in arsenic cancer risk assessment. Environ. Health Perspect. 103(7-8):684-689.

Newman, P.E. 1999. Can reduced folic acid and vitamin B12 levels cause deficient DNA methylation producing mutations which initiate atherosclerosis? Med. Hypotheses 53(5):421-424.

NRC (National Research Council). 1982. Diet, Nutrition and Cancer. Washington, DC: National Academy Press.

NRC (National Research Council). 1993. Pesticides in the Diets of Infants and Children. Washington, DC: National Academy Press.

NRC (National Research Council). 1994. Science and Judgment in Risk Assessment. Washington, DC: National Academy Press.

NRC (National Research Council) 1996. Understanding Risk: Informing Decisions in a Democratic Society. Washington, DC: National Academy Press.

NRC (National Research Council). 1999. Arsenic in Drinking Water. Washington, DC: National Academy Press.

The Presidential/Congressional Commission on Risk Assessment and Risk Management. 1997. Framework for Environmental Health Risk Management. Final Report. Vol. 1. Washington, DC: The Commission.

Ruan, Y., M.H. Peterson, E.M. Wauson, J.G. Waes, R.H. Finnell, and R.L. Vorce. 2000. Folic acid protects SWV/Fnn embryo fibroblasts against arsenic toxicity. Toxicol. Lett. 117(3):129-137.

Schoof, R.A., L.J. Yost, J. Eickhoff, E.A. Crecelius, D.W. Cragin, D.M. Meacher, and D.B. Menzel. 1999. A market basket survey of inorganic arsenic in food. Food Chem. Toxicol. 37(8):839-846.

Smith, A.H., A.P. Arroyo, D.N. Mazumder, M.J. Kosnett, A.L. Hernandez, M. Beeris, M.M. Smith, and L.E. Moore. 2000. Arsenic-induced skin lesions among Atacameno people in Northern Chile despite good nutrition and centuries of exposure. Environ. Health Perspect. 108(7):617-620.

Smith, A.H., M. Goycolea, R. Haque, and M.L. Biggs. 1998. Marked increase in bladder and lung cancer mortality in a region of Northern Chile due to arsenic in drinking water. Am. J. Epidemiol. 147(7):660-669.

Smith, A.H., C. Hopenhayn-Rich, M.N. Bates, H.M. Goeden, I. Hertz-Picciotto, H.M. Duggan, R. Wood, M.J. Kosnett, and M.T. Smith. 1992. Cancer risks from arsenic in drinking water. Environ. Health Perspect. 97:259-267.

Steinmetz, K.A., and J.D. Potter. 1991a. Vegetables, fruit, and cancer. I. Epidemiology. Cancer Causes Control 2(6):325-357.

Steinmetz, K.A., and J.D. Potter. 1991b. Vegetables, fruit, and cancer. II. Mechanisms. Cancer Causes Control 2(6):427-442.

Styblo, M., and D.J. Thomas. 2001. Selenium modifies the metabolism and toxicity of arsenic in primary rat hepatocytes. Toxicol. Appl. Pharmacol. 172(1):52-61.

Styblo, M., L.M. Del Razo, E.L. LeCluyse, G.A. Hamilton, C. Wang, W.R. Cullen, and D.J. Thomas. 1999. Metabolism of arsenic in primary cultures of human and rat hepatocytes. Chem. Res. Toxicol. 12(7):560-565.

Styblo, M., L.M. Del Razo, L. Vega, D.R. Germolec, E.L. LeCluyse, G.A. Hamilton, W. Reed, C. Wang, W.R. Cullen, and D.J. Thomas. 2000. Comparative toxicity of trivalent and pentavalent inorganic and methylated arsenicals in rat and human cells. Arch. Toxicol. 74(6):289-299.

Su, J.H., Y.L. Guo, M.D. Lai, J.D. Huang, Y. Cheng, and D.C. Christiani. 1998. The NAT2* slow acetylator genotype is associated with bladder cancer in Taiwanese, but not in the Black Foot Disease endemic area population. Pharmacogenetics 8(2):187-190.

Thompson, K.M., and J.D. Graham. 1996. Going beyond the single number: using probabilistic risk assessment to improve risk management. Hum. Ecol. Risk Asses. 2(4):1008-1034.

Tao, S.S., and P.M. Bolger. 1999. Dietary arsenic intakes in the United States: FDA Total Diet Study, September 1991-December 1996. Food Addit. Contam. 16(11):465-472.

Tsai, S.M., T.N. Wang, and Y.C. Ko. 1999. Mortality for certain diseases in areas with high levels of arsenic in drinking water. Arch. Environ. Health 54(3):186-193.

Tseng, W.P. 1989. Blackfoot disease in Taiwan: a 30-year follow-up study. Angiology 40(6):547-558.

Tseng, W.P., H.M. Chu, S.W. How, J.M. Fong, C.S. Lin, and S. Yeh. 1968. Prevalence of skin cancer in an endemic area of chronic arsenicism in Taiwan. J. Natl. Cancer Inst. 40(3):453-463.

Tsuda, T., A. Babazono, E. Yamamoto, N. Kurumatani, Y. Mino, T. Ogawa, Y. Kishi, and H. Aoyama. 1995. Ingested arsenic and internal cancer: a historical cohort study followed for 33 years. Am. J. Epidemiol. 141(3):198-209.

Vahter, M. 1999a. Methylation of inorganic arsenic in different mammalian species and population groups. Sci. Prog. 82(Pt 1):69-88.

Vahter, M. 1999b. Variation in human metabolism of arsenic. Pp. 267-279 in Arsenic Exposure and Health Effects, W.R. Chappell, C.O. Abernathy, and R.L. Calderon, eds. Oxford: Elsevier.

Vahter, M. 2000. Genetic polymorphism in the biotransformation of inorganic arsenic and its role in toxicity. Toxicol. Lett. 112-113:209-217.

Vahter, M. 2001. Mechanisms of arsenic biotransformation. Toxicology. In Press.

Valentine, J.L., H.K. Kang, and G. Spivey. 1979. Arsenic levels in human blood, urine, and hair in response to exposure via drinking water. Environ. Res. 20(1):24-32.

Warner, M.L., L.E. Moore, M.T. Smith, D.A. Kalman, E. Fanning, and A.H. Smith. 1994. Increased micronuclei in exfoliated bladder cells of individuals who chronically ingest arsenic-contaminated water in Nevada. Cancer Epidemiol. Biomarkers Prev. 3(7):583-590.

Weinshilboum, R.M., D.M. Otterness, and C.L. Szumlanski. 1999. Methylation pharmacogenetics: catecol O-methyltransferase, thiopurine methyltransferase, and histamine *N*-methyltransferase. Annu. Rev. Pharmacol. Toxicol. 39:19-52.

Willett, W. 1990. Nutritional Epidemiology. New York: Oxford University Press.

World Cancer Research Fund Panel. 1997. Food, Nutrition and the Prevention of Cancer: A Global Perspective. Washington, DC: World Cancer Research Fund Panel, American Institute for Cancer Research.

Wu, M.M., T.L. Kuo, Y.H. Hwang, and C.J. Chen. 1989. Dose-response relation between arsenic concentration in well water and mortality from cancers and vascular diseases. Am. J. Epidemiol. 130(6):1123-1132.

Yu, R.C., K.H. Hsu, C.J. Chen, and J.R. Froines. 2000. Arsenic methylation capacity and skin cancer. Cancer Epidemiol. Biomarkers Prev. 9(11):1259-1262.

5

Quantitative Assessment of Risks Using Modeling Approaches

OVERVIEW OF THE SCIENCE UNDERLYING EPA'S 2001 PROPOSED REGULATION

On January 22, 2001, following the publication of a proposed rule for arsenic in drinking water (EPA 2000a) and a period of public comment, EPA published a final rule for arsenic in drinking water in the *Federal Register*, setting a maximum contaminant level goal (MLCG) of zero for arsenic in drinking water and a maximum contaminant level (MCL) for arsenic of 10 μg in drinking water (EPA 2001). Typically, when developing an MCLG and an MCL, a risk assessment is conducted. Two important components of a risk assessment are hazard identification and dose-response assessment (NRC 1983). Exposure assessment and risk characterization are also important steps in a risk assessment, but they are beyond the scope of this subcommittee's charge and, therefore, will not be discussed here. The purpose of hazard identification is to determine whether the agent in question causes adverse effects. Deciding which end point is the most sensitive and which studies or data sets are most appropriate for assessing the risks from a chemical are major conclusions from a hazard identification. In the case of EPA's assessment of arsenic, the risk being assessed is the risk to the U.S. population from consumption of arsenic in drinking water. The purpose of dose-response assessment is to determine the relationship between the dose and the incidence of an adverse effect in humans. Major conclusions from the dose-response assessment include the model or models that can be used to best determine the risks to the U.S. population from arsenic in drinking water and understanding of the impacts of different model choices on the risk estimates from that analysis. The details of EPA's hazard identification (choice of end point and choice of study) and dose-response assessment (choice of model, selection of a comparison group, and adjustments for water intake, diet, and mortality versus incidence) are discussed below.

Hazard Identification

Choice of End Point

EPA's hazard analysis is included in Section III of the proposed rule (EPA 2000a) and in Section III.D.1 of the final rule (EPA 2001). EPA "relied upon the NRC [1999] report as presenting the best available, peer reviewed science as of its completion and has augmented it with more recently published, peer reviewed information" in its proposed rule. EPA (2000a) concludes that acute or short-term effects are not seen at 50 µg/L (the MCL at the time of the proposal) and, therefore, addresses the "long-term, chronic effects of exposure to low concentrations of inorganic arsenic in drinking water."

With respect to long-term effects, EPA concludes that "arsenic is a multi-site human carcinogen by the drinking water route," and on the basis of epidemiological studies of Asian, Mexican, and South American populations, those "with exposures to arsenic in drinking water generally at or above several hundred micrograms per liter are reported to have increased risks of skin, bladder, and lung cancer." EPA also notes that increased risk of liver and kidney cancer have been associated with arsenic exposure and that skin cancer has been associated with inorganic arsenic contamination in Argentina (reviewed by Neubauer 1947, as cited in EPA 2000a), in Poland (EPA 2000a), and in a dose-dependent manner following exposure to arsenic in drinking water in Taiwan (Tseng et al. 1968, Tseng 1977). Other epidemiological studies also support an association between arsenic exposure and skin cancer (Roth 1956; Albores et al. 1979; Cuzick et al. 1982; Cebrian et al. 1983).

EPA discussed data from two studies in Taiwan demonstrating a statistically significant increase in mortality risks for bladder, kidney, lung, liver, and colon cancer (Chen et al. 1985), and a significant dose-response relationship for death from bladder, kidney, skin, and lung cancer in both sexes and from liver and prostate cancer in males (Wu et al. 1989). An increase in internal cancers was also seen in Argentina (bladder, lung, and kidney cancer) (Hopenhayn-Rich et al. 1996, 1998) and in Chile (bladder, kidney, and lung cancer) (Smith et al. 1998). Tsai et al. (1999) reported that lung, bladder, intestinal, rectal, and laryngeal cancer were associated with chronic exposure to arsenic in drinking water in Taiwan. EPA also reviewed a study by Lewis et al. (1999) that reported mortality of a population in Utah exposed to lower concentrations (average, 18-191 µg/L) of arsenic in drinking water in which there was a statistically significant increase in prostate-cancer mortality, but no increase in bladder or lung cancer mortality. EPA also discussed a study by Kurttio et al. (1999) that found a significant association in a case-control study in Finland between bladder cancer and exposure to very low concentrations of arsenic in drinking water (odds ratio of 1.53, 95% confidence interval (CI) = 0.75-3.09 at 0.1-0.5 µg/L; 2.44, 95% CI = 1.11-5.37 at greater than 0.5 µg/L). No association was seen for kidney cancer.

EPA reviewed noncancer effects that are observed following chronic exposure to arsenic including dermal effects (Yeh 1973; Tseng 1977; Cuzick et al. 1982), gastrointestinal effects (Morris et al. 1974; Nevens et al. 1990; Mazumder et al. 1997), peripheral vascular disease (Tseng 1977; Zaldivar 1974; Cebrian 1987; Lewis et al. 1999), and diabetes (Lai et al. 1994; Rahman and Axelson 1995; Rahman et al. 1998).

In the final rule, EPA again summarized the acute and chronic effects of arsenic, and added a discussion of a study from Japan (Tsuda et al. 1995). The study found an association between exposure to arsenic in drinking water and lung and bladder cancer. In addition, EPA (2001) added a short discussion on the potential susceptibility of children to arsenic. EPA agreed with the conclusion of the majority of the EPA Science Advisory Board (SAB) members (EPA 2000b) that children are generally at greater risk for a toxic response to any agent in water because of their greater drinking-water consumption (on a unit-body-weight basis), but that the available data, including a study of infant mortality in Chile (Hopenhayn-Rich et al. 2000), do not demonstrate a heightened susceptibility to arsenic.

After discussing all the toxic effects of arsenic, the water concentrations at which they occur, and the NRC (1999) report, EPA chose cancer as the most sensitive end point, stating that it "focused its risk assessment on the carcinogenic effects of inorganic arsenic" (EPA 2000a). EPA (2001) states that lung and bladder cancer are the internal cancers most consistently seen and best characterized in epidemiological studies, and its quantitative risk assessment is based on data for those two cancers.

Choice of Study

An important decision in a quantitative risk assessment is the choice of critical study (or studies) to be used in the dose-response assessment. At the time of EPA's proposed rule, few animal carcinogenicity bioassays had been conducted for arsenic, and there were no positive animal models for dose-response modeling. There was, however, a "large data base on the effects of arsenic on humans" (EPA 2000a, p. 38902). EPA concluded that questions remain about the shape of the dose-response relationship at low concentrations. The advantages of using the studies from southwestern Taiwan (Chen et al. 1985; Wu et al. 1989) for quantitative risk assessment, according to EPA, are the duration of exposure and follow-up, the size of the population (more than 40,000 individuals), the extensive pathology data, and the homogenous lifestyles of the population. Those studies are limited, however, by their design (i.e., they are ecological epidemiology studies), which makes quantitative evaluation of dose-response relationships more difficult. EPA also stated that the studies from Chile (Smith et al. 1998) and Argentina (Hopenhayn-Rich et al. 1996; 1998) are more limited than the Taiwanese studies (Chen et al. 1985; Wu et al. 1989) and not suitable for quantitative dose-response assessment, but that they provide supportive evidence for the effects seen in southwestern Taiwan. EPA concluded that "[t]hese epidemiological studies provide the basis for assessing potential risk from lower concentrations of inorganic arsenic in drinking water" (EPA 2000a, p. 38902). In its final rule, EPA also concluded that the Utah study by Lewis et al. (1999) "is not powerful enough to estimate excess risks with enough precision to be useful for the Agency's arsenic risk analysis."

Therefore, in its final rule, EPA (2001) still considered the southwestern Taiwan data to be the critical data set for conducting a quantitative risk assessment for exposure to arsenic in drinking water.

Dose-Response Modeling

Model Choice and Selection of a Comparison Group

In its proposed arsenic rule, EPA concluded, on the basis of the NRC (1999) report, that there is "no basis for determining the shape of a sublinear dose-response curve for inorganic arsenic" (EPA 2000a). Therefore, EPA estimated the risks of cancer from exposure to arsenic in drinking water using a linear extrapolation from the data observed in the southwestern Taiwanese epidemiological studies down to the origin. EPA's default to a linear extrapolation in the absence of adequate mode-of-action data (EPA 1996) is in part a policy decision. For the proposed rule, EPA used the bladder cancer risk estimates presented in the NRC (1999) report (see Table 5-1 for examples). EPA cited a lifetime risk estimate with a 95% upper confidence limit of 1 to 1.347 per 1,000, calculated by a Poisson regression model not using any baseline data (i.e., no comparison group) (NRC 1999), and EPA used four distributions of risk estimates (mean and 95% CI) from NRC (1999) as representative risks in a Monte Carlo analysis to estimate the potential health benefits from the proposed rule. Those four distributions all come from analyses of the southwestern Taiwanese data (Chen et al. 1985; Wu et al. 1989) using a Poisson regression model with age entered as a quadratic function and dose entered as a linear function, either with or without baseline data, or a Poisson regression model with a point-of-departure approach, with or without baseline data.

On October 20, 2000, EPA published a Notice of Data Availability in the *Federal Register* (EPA 2000b) in which it discussed statistical modeling published by Morales et al. (2000) and indicated that those analyses would be considered in its final rule for arsenic in drinking water. Morales et al. (2000) estimated bladder, lung, and liver cancer risks for the southwestern Taiwanese population based on the same data set that was analyzed by NRC (1999). Morales et al. (2000) calculated cancer risk estimates using 10 risk models and also considered how well those models fit the data sets. Of those models, EPA chose a single model that did not use an external comparison population either from all of Taiwan or part of southwestern Taiwan, because most of the models that incorporate a comparison population result in a dose-response curve that is supralinear at low doses. EPA indicated that there is no biological basis for a supralinear curve. In addition, differences other than arsenic exposure between the study population and the comparison population could affect the results. The decision to use a dose-response model that does not incorporate a comparison population agreed with the SAB's recommendation that the analysis should be conducted without a comparison group (see discussion below). Of the models that did not incorporate a comparison population, model 1 from Morales et al. (2000), in which "the relative risk of mortality at any time is assumed to increase exponentially, with a linear function of dose and a quadratic function of age," was used because it best fit the data based on the Akaike information criterion (EPA 2001). However, EPA did not

TABLE 5-1 NRC's Risk Estimates for Bladder Cancer Mortality from 1999 NRC Report[a]

Method of Analysis	Point of Departure, μg/L[b]	Risk at 50 μg/L (× 1,000)	Margin of Exposure at 50 μg/L[c]
Poisson model, linear dose, no background data	404 (323)	1.237 (1.548)	8.08 (6.46)
Poisson model, linear dose, background data included	443 (407)	1.129 (1.229)	8.86 (8.14)

[a] Estimated points of departure at the 1% excess risk level, corresponding margin of exposure at 50 μg/L and corresponding excess lifetime risk estimates at 50 μg/L for bladder cancer in males. Figures in parentheses are 95% confidence limits (lower for the point-of-departure estimates; upper for estimated risk at 50 μg/L). Risk estimates are those predicted in Taiwan using U.S. ingestion rates.
[b] The point of departure represents an estimate or observed level of exposure or dose associated with an increase in adverse effects in the study population. An example of a point of departure is an ED_{01}.
[c] A margin-of-exposure analysis compares the levels of arsenic to which the U.S. population is exposed with the point of departure to characterize the risk to the U.S. population. The larger the ratio, the greater degree of assumed safety for the population.
Abbreviation: ED_{01}, 1% effective dose.
Source: Modified from NRC (1999).

publish the theoretical risk estimates on which it based its analyses in the *Federal Register* (EPA 2000a, 2001). The risk estimates that it presents (EPA 2001) are adjusted for the occurrence of arsenic in U.S. drinking water; consideration of such an adjustment is beyond the charge to this subcommittee. Because EPA did not present theoretical lifetime excess bladder or lung cancer risk estimates, the subcommittee used linear extrapolation from the ED_{01}s presented in Morales et al. (2000) to estimate these risks at 3, 5, 10, and 20 μg/L (Table 5-2).

Adjustments for Water Intake

To estimate cancer risks associated with a given arsenic concentration in drinking water, assumptions must be made about water consumption in both

TABLE 5-2 Theoretical Estimates of Excess Lifetime Risk (Incidence per 10,000 people) of Lung and Bladder Cancer at Various Concentrations of Arsenic in Drinking Water Based on ED_{01} Values from Morales et al. (2000)[a]

Arsenic Concentration (μg/L)	Bladder Cancer		Lung Cancer	
	Females	Males	Females	Males
3	1.2	.76	1.2	0.82
5	2.0	1.3	1.9	1.4
10	4.0	2.5	3.9	2.7
20	7.9	5.1	7.8	5.5

[a] Excess cancer risk estimates were calculated using the ED_{01}s estimated by Morales et al. (2000) using a model in which the relative risk of mortality at any time is assumed to increase exponentially, with a linear function of dose and a quadratic function of age (i.e., a multiplicative Poisson linear regression); no external comparison population was used (see Model 1 of Table 8 from Morales et al. 2000). Risk estimates are rounded to two significant figures. The Taiwanese exposure per kilogram of body weight is assumed to be 2.2 times the U.S. exposure.

the U.S. population and the study population. EPA estimated mean daily average per capita consumption of water by individuals in the United States is 1 L/person/day for "community tap water" and 1.2 L/person/day for "total water" (which includes bottled water) based on data from the 1994-1996 Continuing Survey of Food Intakes by Individuals (CSFII) (EPA 2000d). The 90th percentile is 2.1 L/person/day and 2.3 L/person/day for community tap water and total water, respectively. Rather than only using a point estimate for its risk assessment, EPA conducted a Monte Carlo analysis using the CFSII data to incorporate water intake. Those distributions take into account age, sex, and weight. EPA assumed that the Taiwanese consumed relatively more water per unit of body weight than Americans, estimating consumption of 3.5 and 2.0 L/day for men and women in Taiwan, respectively. As discussed in the following section, EPA also added water consumption to account for water used in cooking in Taiwan. It should be noted that assumptions that increase the amount of arsenic consumed (drinking water and diet) by the study population reduce the "potency factor" or estimated slope of the linear dose-response function when applied to other populations, thereby decreasing the estimated risk in other populations. Conversely, underestimation of the actual arsenic intake in the study population increases risk estimates in other populations. Therefore, assumptions about total arsenic exposure in the study population can have a large impact on risk estimates.

Adjustments for Dietary Intake of Arsenic

The staple foods in the southwestern Taiwan region where the study population resided were rice and sweet potatoes. Those foods absorb a great deal of water when cooked. As part of its risk assessment of arsenic in drinking water, EPA (2001) adjusted its lower-bound estimates to account for exposure to arsenic in food from cooking water. For that adjustment, EPA assumed that people in the study population eat 1 cup of dry rice and 2 pounds of sweet potatoes per day. To adjust for arsenic absorbed during cooking, EPA added 1 L of water consumption to the Taiwanese population. Therefore, in its analyses, EPA assumed that Taiwanese men and women consumed the equivalent of a total of 4.5 L/day and 3.0 L/day of water, respectively. Although EPA used a Monte Carlo analysis to account for variability in U.S. water consumption rates, its analyses did not incorporate analogous variability in the Taiwanese water consumption rates.

EPA also discussed the fact that the food in Taiwan contains more arsenic than the food in the U.S., even prior to cooking. NRC (1999) presented data indicating that individuals in Taiwan consume food containing inorganic arsenic at 50 µg/day, compared with 10 µg/day for Americans. To account for the intake of arsenic from food, EPA multiplied the lower-bound risk estimates by the fraction of arsenic consumed per kilogram contributed by drinking water (calculated by dividing the arsenic ingested from drinking water (µg/kg/day) by the total arsenic consumed from drinking water, cooking water, and food) (J. Bennett, EPA, personal commun. May 22, 2001).

Adjustments for Mortality Versus Incidence

EPA's dose-response assessment is derived from data on mortality from bladder and lung cancer in the Taiwanese study (Chen et al. 1985, 1992; Wu et al. 1989). Extrapolating the mortality-risk estimates calculated in the Taiwanese population to the incident risks in the U.S. population requires an adjustment for the survival rate for bladder and lung cancer. EPA (2001) noted that the Taiwanese people in the study population had low incomes and poor diets, and that "the availability and quality of medical care is not of high quality, by U.S. standards." Therefore, EPA assumed that the bladder cancer incidence was relatively close to the bladder cancer mortality in the Taiwanese study area. EPA calculated the survival rate for bladder cancer by considering the survival-rate data compiled by the World Health Organization (WHO) for bladder cancer in developing countries from 1982 through 1992 (IARC 1999) and by comparing the annual bladder cancer mortality and incidence for the general population of Taiwan in 1996. From those data, EPA concluded that "bladder cancer incidence could be no more than two-fold bladder cancer mortality; and that an 80% mortality rate would be plausible" (EPA 2001; page 7009). Therefore, when calculating the bladder cancer cases avoided at a given MCL, EPA adjusted the upper bound by a factor of 1.25 to reflect the mortality for bladder cancer. With respect to lung cancer, EPA concluded that "because lung cancer [mortality] rates are quite high, about 88% in the U.S. [EPA 1998], the assumption was made that all lung cancers in the Taiwan study area resulted in fatalities."

OVERVIEW OF THE SAB'S REPORT ON THE 2001 RISK ASSESSMENT

EPA charged the SAB to review the proposed arsenic rule (EPA 2000a) and to specifically review (1) EPA's focus on inorganic arsenic as the principle form of arsenic causing health effects; (2) the implications of exposure to natural arsenic through food; and (3) the appropriateness of EPA's precautionary advice to use low-arsenic water in the preparation of infant formula. They also requested the SAB to address several questions related to treatment options for arsenic in drinking water.

On December 12, 2000, the SAB issued a report on the proposed drinking-water regulation (EPA 2000c), responding to those questions on the scientific basis of EPA's health risk assessment and on the economic and engineering aspects of the final rule. The exposure assessment, costs, benefits, control technologies, and policy issues discussed by the SAB are beyond the charge to this NRC subcommittee and will not be discussed. The SAB's responses to EPA's three questions on the health effects of arsenic are discussed in this section. EPA addressed some of the SAB's comments in its final arsenic rule (EPA 2001).

Inorganic Arsenic As Principal Form of Arsenic Causing Health Effects

The SAB pointed out that new data released since the 1999 NRC report indicated that inorganic forms of arsenic are not solely responsible for the toxic effects of arsenic (see Chapter 3 for discussion of new data). Because exposures to other forms of arsenic can produce health effects, the SAB recommended that future risk assessments provide quantitative information on how the intake of inorganic arsenic is related to the concentration of arsenic metabolites in the urine and to bladder cancer. However, because the principal forms of arsenic in drinking water are inorganic, the SAB believed "that it is appropriate for the Agency [EPA] to make [inorganic arsenic] its regulatory focus."

Implications of Exposure to Natural Arsenic Through Food

The SAB concluded that on average, for the general U.S. population, ingestion of inorganic arsenic via food was considerably greater than ingestion of inorganic arsenic via drinking water. The SAB agreed that data were not available to determine "a well-defined nonlinear dose-response curve." Furthermore, the SAB concluded that insufficient data on the distribution of food intakes existed to adequately consider them in the analysis. Therefore, the SAB concluded that EPA had no choice but to calculate marginal risk reductions based solely on arsenic concentrations in drinking water. The SAB noted that "there is a limit to the benefits that can be realized by reducing arsenic in drinking water" as long as the concentrations in food remain unchanged. The SAB also reiterated the recommendation in the NRC (1999) report to obtain additional studies on the noncancer effects of arsenic and incorporate that information into a risk-assessment and cost-benefit analysis.

Health Advisory on Low-Arsenic Water and Infant Formula

The SAB also reviewed EPA's plan to issue a health advisory for the use of low-arsenic water in the preparation of infant formula. Some of SAB's responses focused on risk communication and the public's ability to follow such an advisory; those comments are beyond the scope of this subcommittee's charge. However, the SAB also discussed the issue of children's susceptibility to arsenic. Most of the SAB members agreed that special circumstances make infants unique in regard to their response to drinking-water contaminants, but the SAB did not reach consensus on endorsing EPA's intent to issue a health advisory.

An SAB consultant wrote and one SAB member endorsed a minority report on the issue of infant and children's risk. The minority report stated that differences in respiratory and circulatory flow rates, cell-proliferation rates, enzymatic pathways, developmental process, life expectancies, and the disposition of chemicals in infants and other differences in cells during development can make children more susceptible to toxic chemicals, including arsenic. It further concluded that data from Hopenhayn-Rich et al. (2000) and Concha et al. (1998) "indicate that young children are a uniquely sensitive population for adverse health effects of arsenic." The minority report "departs from the majority opinion contained in the [SAB] report in strength of its conclusions if not the general reasonableness of the need for increased concern for children, which is also held by the [SAB]." The basis for increased concern for children is uncertainty about pulmonary and cardiovascular risks to infants, high exposure of infants on a per kilogram basis, and the longer period of exposure and outcome relative to adults. The latter is particularly relevant if the latency period for cancer development from low arsenic exposure is long or if the appropriate dose metric involves a less-than-lifetime exposure as discussed in Chapter 4.

Although the SAB recognized that children differ from adults in many ways that could make them more susceptible to toxic chemicals, "the majority of the [SAB] did not feel that data available to them on arsenic had demonstrated an increased sensitivity to arsenic in children." The SAB concluded that available data on U.S. drinking-water consumption indicate that infants who consume formula made from drinking water could have a higher dose of arsenic per unit of body weight than adults. The majority of the SAB did not believe that the study by Hopenhayn-Rich et al. (2000) demonstrated "a heightened sensitivity or susceptibility to arsenic" but that the study "appears to be a hypothesis generating study that, in light of the limitations [the SAB described], merits and requires further study before drawing final conclusions."

SAB's Comments on EPA's Interpretation of the NRC Report

The SAB also commented on EPA's interpretation of the 1999 NRC report. The SAB, in general, agreed with the 1999 NRC report, which formed part of the basis for EPA's proposed regulation. The SAB reiterated some of the cautions made in the 1999 NRC report surrounding the use of the ecological Taiwanese data to characterize risks to the U.S. populations, including (1) the potential for measurement error and confounding because of uncertainty in exposure assessment; (2) the impact of the choice of model; and (3) other factors "such as poor nutrition and low selenium concentrations in Taiwan, genetic and cultural characteristics, and arsenic intake from food." The SAB stated that EPA "may have taken the modeling activity in the NRC report as prescriptive." It cited a study by Morales et al. (2000), published after EPA published its proposed regulation (EPA 2000a), which could impact its risk assessment. The SAB also concluded "that the comparison populations [used in some analyses] were not appropriate control groups for the study area" in the Taiwanese study and, therefore, recommended dose-response models that did not use the comparison population. Following this recommendation, EPA used the Morales et al. (2000) study and analyses conducted without the use of a comparison population in its final rule (EPA 2001).[1]

The SAB did not believe, however, that "resolution of all these factors can nor must be accomplished before EPA promulgates a final arsenic rule." Furthermore, the SAB agreed that the "available data do not yet meet EPA's new criteria for departing from linear extrapolation of cancer risk."

THE SUBCOMMITTEE'S EVALUATION

This section critiques various elements that influenced EPA's decision to propose a new MCL for arsenic of 10 µg/L (ppb) of drinking water based on what was known at that time and on new information that has emerged since publication of the 1999 NRC report. Issues to be addressed include choice of end point (bladder and lung cancer), and the use of the southwestern Taiwanese study as the basis for EPA's risk assessment. This section will also critique and discuss a number of the statistical modeling assumptions used by EPA, including (1) the inclusion of an external comparison population in the analysis, (2) the choice of dose-response model, and (3) the approach used to

---

[1] The SAB report stated that it "focused on evidence provided in the NRC 1999 report that indicated that the population in the study area differed substantially from these comparison populations socioeconomically and in diet." (EPA 2000c, p. 26). However, it did not specifically identify any part of the NRC report that compared nutritional or socioeconomic status between the study population and the southwestern Taiwanese referent population.

address the impact of assumptions about drinking-water intake in the study population.

Choice of End Point and Study

EPA's decision to base its risk assessment on the bladder and lung cancer mortality data from the southwestern Taiwanese studies (Chen et al. 1985, 1988, 1992; Wu et al. 1989, as analyzed by Morales et al. 2000) was largely in response to the recommendations in the 1999 NRC report. Arsenic has been implicated in a variety of adverse health effects, including dermal effects, gastrointestinal effects, peripheral vascular disease, diabetes, and cancer. Cancer, however, is the end point most consistently seen as a consequence of long-term chronic exposure. It was also the end point with the most extensive quantitative information available on dose-response. Although several studies that were available at the time of the 1999 NRC report convincingly confirmed the carcinogenic effects of arsenic (e.g., Hopenhayn-Rich et al. 1996, 1998; Smith et al. 1998), the southwestern Taiwanese studies documenting bladder and lung cancer (Chen et al. 1985, 1988, 1992; Wu et al. 1989) were the only ones that quantified exposure levels well enough to support a quantitative dose-response analysis. Since the 1999 NRC report, considerable new information has emerged related to the health effects of arsenic in drinking water. Cancer, however, remains the end point with the most reliable data available for the purpose of quantifying dose-response effects (see Chapter 2). Although the southwestern Taiwanese studies still stand out as the strongest sources of dose-response information, two recently published studies (Ferreccio et al. 2000; Chiou et al. 2001) are of sufficient quality to warrant serious consideration as part of the quantitative risk assessment for arsenic in drinking water. Both studies are described and critiqued in detail in Chapter 2. They are summarized briefly here, and their potential for use in a quantitative risk assessment is also discussed.

In a prospective cohort study, Chiou et al. (2001) assessed arsenic exposure at the individual level for over 8,000 subjects in northeastern Taiwan. Drinking-water arsenic concentrations ranged from below detection to 3,843 µg/L of drinking water, with the majority of study subjects (75%) exposed at concentrations less than 100 µg/L. More than half of the subjects were exposed at drinking-water concentrations less than 50 µg/L, and 2,346 consumed drinking water with no detectable arsenic (less than 0.15 µg/L). Data are also available at the individual level on potential confounding factors, such as smoking and socioeconomic status. The incidences of urinary-tract cancer overall and the more specific urinary-tract transitional-cell carcinoma (TCC) are the end points of interest. Although the study involves a relatively large number of subjects, because of the relatively short follow-up period (approximately 4 years), the number of cancers observed is relatively low (18 urinary cancers, of which 11 were classified as TCC).

Ferreccio et al. (2000) conducted a case-control study in an area of northern Chile, where 151 lung cancer cases and 419 controls were studied. Average exposures to arsenic during the period 1930 to 1994 were grouped as 0-10 µg/L (referent group), 10-29 µg/L, 30-39 µg/L, 40-49 µg/L, 50-199 µg/L, and 200-400 µg/L. There is individual information on residential history, socioeconomic status, occupational history, and smoking. Another strength of that study is that it has detailed, individual-specific exposure assessment; arsenic exposure was determined for each subject by residence and historical arsenic concentrations on a yearly basis from 1930 to 1994. Drinking-water arsenic concentrations were very high (860 µg/L) in the major city of the study area, Antofagasta, from 1958 to 1970. Risk calculations based on exposure during this period were somewhat lower than risk calculations based on average exposure from 1930 to 1994.

Quantitative Risk Estimates

The usefulness of the Chiou et al. study to quantitative risk analysis are limited by its follow-up period compared with that of the studies from southwestern Taiwan (Chen et al. 1985, 1992; Wu et al. 1989). It is useful, however, to consider a qualitative synthesis of the results from the different studies and end points to assess the concordance of the estimates of the effective doses under different modeling assumptions. In this section, 1% effective dose $(ED_{01})^2$ calculations for the two recent studies are discussed (Table 5-3). In the next section, some of the modeling assumptions and choices that can influence ED calculations are described, and finally, a summary of the subcommittee's findings is presented.

EPA based its risk assessment on ED_{01}s calculated by the NRC (1999), and also by Morales et al. (2000). The ED_{01}s were based on Poisson regression models. Although simpler methods could also be applied, the subcom-

---

[2]The 1% effective dose is the concentration of arsenic in drinking water in the study that is associated with a 1% increase in the excess risk of cancer.

mittee thinks that the general Poisson approach is appropriate. Poisson regression does not impose any strong model assumptions. It is simply a probabilistic formulation for handling the analysis of rare events. As such, it is appropriate for the analysis of cancer data collected in a cohort study setting (see Breslow and Day (1988) for related discussion). Modeling assumptions play a critical role, however, when it comes to the choice of the parametric form relating covariates to the rate parameter of interest (in the present context, cancer mortality). Although the standard formulation of Poisson regression involves modeling the rate parameter of interest as an exponential function of covariates, other formulations are possible with specialized programing.

In the following sections, the subcommittee considers models that characterize cancer mortality as a baseline function of age (and possibly other covariates) multiplied by a relative function, $g(d)$, where d is dose. The subcommittee considers two models of relative risk. One of these, the multiplicative model, assumes that the logarithm of $g(d)$ can be expressed as a linear function of exposure. This model can be fit using standard statistical software for Poisson regression. The second formulation, the additive model, assumes that $g(d)$ is simply a linear function of dose. (This formulation requires specialized programing.) Although a variety of other formulations could be considered, relative risk models have a lengthy precedent in cancer epidemiology and there are good reasons to believe that relative risks are similar across populations, even when baseline risks vary or synergy is operating (e.g., as might possibly be the case for smoking and arsenic exposure).

The formula used by the NRC (1999; also see Morales et al. 2000) for computing an ED_{01} based on the results of the Poisson regression analysis had some limitations; the primary one being that it was not straightforward to incorporate the baseline cancer risk based on the U.S. population. A better approach is to base the ED calculation on the formula presented in Appendix II of the BEIR IV analysis of lung cancer mortality associated with radon exposure (NRC 1988). Given a specified simple functional form (e.g., linear) for the relative mortality risk associated with exposure to arsenic in drinking water, the formula can be used to compute a lifetime risk estimate, and hence an ED, based only on the value of the relative risk at a single specified value of exposure (Gail 1975). This property is particularly advantageous for computing EDs in settings where raw data are unavailable but where estimates of relative risks and their associated confidence intervals (CIs) have been published (e.g., lung cancer odds ratios reported by Ferreccio et al. 2000). The BEIR IV formula can be applied with either incidence or mortality. In the following section, the application of BEIR IV using mortality is considered; later in the chapter, its use with incidence is considered.

TABLE 5-3 1% Effective Doses Calculated for Different Studies Under Different Modeling Assumptions[a]

Study	Site	Modeling Details	Risk Ratio at 50 ppb (95% CI)	ED_{01} (LED_{01})[a], µg/L
Ferreccio et al. 2000[b]	Lung	Linear regression applied to relative risks in published paper, using average arsenic concentrations from 1930 to 1994	2.4 (1.9, 2.9)[c]	5 (3); male 7 (5); female
	Lung	Linear regression applied to relative risks in published paper, using average arsenic concentrations during peak years from 1958 to 1970	1.4 (1.3, 1.5)[d]	17 (14); male 27 (21); female
Chiou et al. 2001[b]	Urinary	Multiplicative, linear dose	1.05 (1.01, 1.09)	500+ (500+); male 500+ (500+); female
		Multiplicative, log dose	1.12 (1.04, 1.20)	500+ (372); male 500+ (500+); female
		Additive, linear dose	1.44 (0.63, 2.24)	229 (80); male 499 (174); female
		Multiplicative, linear dose (restricted to <400 ppb)	1.21 (0.89, 1.64)	326 (125); male 519 (198); female
		Multiplicative, log dose (<400 ppb)	1.25 (0.88, 1.76)	305 (111); male 500+ (181); female
		Additive, linear dose (<400 ppb)	1.47 (0.58, 2.36)	214 (73); male 464 (160); female
		Multiplicative, linear dose (<200 ppb)	1.536 (0.811, 2.91)	144 (58); male 230 (92); female
		Multiplicative, log dose (<200 ppb)	1.57 (0.80, 3.08)	141 (54); male 231 (88); female
		Additive, linear dose (<200 ppb)	1.77 (0.21, 3.34)	129 (42); male 281 (92); female

Study	Cancer	Model	ED$_{01}$ (LED$_{01}$)	Estimate; sex
Chen et al. 1985, 1992	Lung	Multiplicative, linear dose	1.15 (1.10, 1.14)	84 (72); male
		Multiplicative, log dose	1.15 (1.13, 1.18)	65 (56); male
		Additive, linear dose	1.26 (1.25, 1.27)	38 (37); male
		Multiplicative, linear dose	1.16 (1.14, 1.18)	94 (84); female
		Multiplicative, log dose	1.21 (1.18, 1.24)	72 (64); female
		Additive, linear dose	1.46 (1.44, 1.49)	33 (31); female
Chen et al. 1985, 1992	Bladder	Multiplicative, linear dose	1.22 (1.19, 1.24)	317 (286); male
		Multiplicative, log dose	1.29 (1.26, 1.33)	259 (232); male
		Additive, linear dose	1.98 (1.92, 2.14)	102 (94); male
		Multiplicative, linear dose	1.25 (1.23, 1.28)	443 (406); female
		Multiplicative, log dose	1.34 (1.31, 1.38)	371 (336); female
		Additive, linear dose	2.57 (2.42, 2.73)	138 (125); female

[a] Estimated ED$_{01}$s are the doses corresponding to a theoretical 1% excess lifetime risk of cancer mortality in the United States. Drinking rates per unit of body weight are assumed to be the same for the U.S., Taiwanese, and Chilean populations. The Chilean and U.S. populations are assumed to have the same average weights, so no drinking rate adjustment is needed to extrapolate results from Chile to the United States. The typical Taiwanese person is assumed to weigh only 50 or 55 kg compared to 70 kg for the typical U.S. person, so a factor of 1.4 is used to extrapolate results from Taiwan to the United States.

[b] This study reports relative risks associated with cancer incidence. The calculations used assume that the same relative risks apply to cancer mortality.

[c] Based on applying linear regression to odds ratios calculated using exposure to arsenic in drinking water based on average arsenic concentrations from 1930 to 1994 by Ferreccio et al. (2000, Table 5). Model forced through 1 at lowest exposure level.

[d] Based on applying linear regression to odds ratios calculated using exposure to arsenic in drinking water based on average arsenic concentrations during the peak years of exposure, 1958-1970, by Ferreccio et al. (2000, Table 6). Model forced through 1 at lowest exposure level.

Abbreviations: ED$_{01}$, 1% effective dose; LED$_{01}$, lower limit on the dose that gives a 1% effect.

BEIR IV

To describe how the BEIR IV formula can be used, it is necessary to specify some notation. Let h_i be defined as the hazard for cancer death in age interval i. Similarly, h_i^* is defined as the corresponding overall hazard of death. We define q_i to be the conditional probability of surviving the ith time interval, given survival to the beginning of the interval,

$$q_i = \exp(-5 \times h_i) \qquad (1)$$

Finally, S_i is the probability of surviving to the beginning of the ith interval,

$$S_i = \prod_{j=1}^{i-1} (1 - q_i) \qquad (2)$$

Then the lifetime cancer mortality risk (R_0) for an unexposed individual is

$$R_0 = \sum_{i \ \varepsilon \ \text{age groups}} h_i / h_i^* S_i (1 - q_i) \qquad (3)$$

The lifetime cancer mortality risk for an individual exposed to an arsenic concentration of d µg/L (R_d) can be obtained by replacing h_i with $h_i g(d)$, where $g(d)$ is the relative risk associated with that exposure concentration, and the overall mortality hazard for an individual exposed at a dose, d, is

$$h_i^* + (g(d) - 1) h_i \qquad (4)$$

The ED is obtained as the solution to $R_d - R_0 = 0.01$. Data on overall U.S. population deaths and cancer-specific deaths were based on the vital statistics records for 1996 obtained from the National Center for Health Statistics (GMWK I table for deaths). Bladder cancers were those corresponding to ICD9 code 188; lung cancers were those corresponding to ICD9 code 162.

As can be seen from the above discussion, the BEIR IV approach uses relative risks from the epidemiological studies and estimates $ED_{01}s$ for the U.S. population using baseline risks for the U.S. population. In contrast, in

NRC (1999) and Morales et al. (2000), where the BEIR IV approach was not used, the ED_{01}s were estimated using Taiwanese baseline risks. That difference in the two approaches affects the resulting ED_{01}s when the baseline risk in the two countries differs. If the baseline risk in the U.S. population is higher than that in the Taiwanese population, then the ED_{01}s determined using the BEIR IV approach would be lower than those determined using the alternate approach. The subcommittee's comparison of recent data on lung cancer incidence in the United States (Ferlay et al. 2001) with lung cancer incidence in Taiwan (You et al. 2001) indicates that lung cancer incidence is approximately three times higher in the United States than in Taiwan for females and two times higher for males. For bladder cancer, the incidence is approximately three times higher in males (Ferlay et al. 2001; You et al. 2001) and two times higher in females (Ferlay et al. 2001; You et al. 2001; C.J. Chen, National Taiwan University, personal commun., August 28, 2001). The different baseline incidences result in a lowering of the ED_{01}s calculated using the BEIR IV approach relative to the ED_{01}s calculated applying the approach used by NRC (1999) and Morales et al. (2000); the corresponding risk estimates are increased.

It is interesting to note that when h_i is small relative to h_i^*, then the approach described above using additive Poisson regression and the BEIR IV approach yields results very similar to those resulting from the approach recommended by Smith et al. (1992). Although not as accurate as the Poisson modeling approach, the approach used by Smith et al. (1992) has the advantage of being simpler and less computationally burdensome. In that approach, given the lifetime cancer mortality in an unexposed population (R_0 in the notation used here), along with the relative risk, $g(d)$, then the ED_{01} is the solution to $R_0 \times (g(d) - 1) = 0.01$. This simple formula, in fact, can be used as an approximate check of the results based on the BEIR IV formula presented in Table 5-3. For its results in Table 5-3, the subcommittee used cancer mortality data reported by the National Center for Health Statistics for 1996 (CDC 2001). Baseline risks (R_0) for lung cancer were 0.076 and 0.046 for males and females, respectively. The corresponding risks for bladder cancer were 0.007 and 0.003. Calculations of cancer risk estimates presented later in the chapter use the simple approximate formula ($R_0 \times (g(d) - 1) = 0.01$) to derive ED_{01}s with respect to lifetime cancer incidence rather than mortality.

Chiou et al. (2001) reported relative risks associated with the incidence of all urinary tract cancers (ICD-9 codes 188 and 189) and specific data for TCC (ICD-9 codes 8120.2, 8120.3, or 8130.3) grouped in several exposure categories, including 0-10.0 µg/L, 10.1-50.0 µg/L, 50.1-100.0 µg/L, and greater than 100.0 µg/L. Treating the 0-10-µg/L group as baseline, the authors report relative risks (based on a multivariate model that included a linear effect of age, as well as a gender and smoking effect) ranging from 1.5 for the 10-50-µg/L group to 15.3 for the greater than 100-µg/L group. The relative risk in the lowest group was not statistically different from 1 and had a very wide CI of 0.3 to 8.0. The authors, however, have made the data available to the subcommittee, allowing exploration of some dose-response models based on the ungrouped exposure variable. Exploratory analysis reveals considerable model sensitivity, which is not surprising given the relatively small number of cancers seen in the short observation time of study (15 urinary tract cancers, including 10 TCCs, in the 4 years for which exposure information was available). The dose-response pattern is quite similar to that seen in the data from southwestern Taiwan, with a somewhat supralinear pattern at low doses and flattening out at high doses. The best fits were provided by a model with a log transformation of dose (including smoking, age, and gender, as described by Chiou et al. 2001) and an additive model with linear dose. As summarized in Table 5-3, ED_{01} estimates based on the additive linear model tended to be higher than those based on the multiplicative model; that pattern is consistent with the bladder cancer results obtained from the southwestern region. When the models were rerun restricting the study to subjects exposed to less than 400 µg/L and again to subjects exposed to less than 200 µg/L, there was a much higher concordance between the various models. The BEIR IV (NRC 1988) formula described above was used to compute ED_{01}s. To apply those ED_{01}s to the U.S. population, the fact that the Chiou et al. (2001) analysis is based on cancer incidence, not mortality, needs to be addressed. For simplicity, it is assumed here that the relative risks are the same for cancer mortality and incidence. That assumption is reasonably accurate in the case of lung cancer, which has a very high case mortality. It might be less accurate for bladder cancer. Assumptions about possible differences between the United States and Taiwan with respect to drinking-water rates are also necessary. It was assumed that the Taiwanese, Chilean, and U.S. populations all drank the same amount of water but that a person in the United States weighs 70 kg and a person in Taiwan or Chile weighs 50 kg. Those assumptions result in a conversion factor of 1.4 between the United States and the other countries (see further discussion below).

Ferreccio et al. (2000) report odds ratios for lung cancer associated with arsenic exposure through public drinking-water supplies for Chile. The authors report their results under several different modeling approaches and choices of control group. The odds ratio associated with the 30-59-μg/L group (based on a multivariate model for the peak-period dose metric that adjusted for age, gender, smoking, and occupational exposure) was 1.8 (95% CI = 0.5-6.9). To obtain a more precise estimate using all the available data, the subcommittee applied a simple linear regression line to the odds ratios listed in Tables 5 and 6 of Ferreccio et al. (2000), using the midpoint of each exposure interval and forcing the line through unity for the lowest exposure group. The analysis yielded an estimated odds ratio of approximately 2.4 (95% CI = 1.9-2.9) based on the data in Table 5 of Ferreccio et al. (2000) and an estimated odds ratio of approximately 1.4 (95% CI = 1.3-1.5) based on the data in Table 6. No adjustments were made for drinking-water intakes and body-weight differences between the Chilean and U.S. populations. The estimated risk ratio of 1.4 translates to an ED_{01} of 17 μg/L for males and 27 for females. Because the values in Table 6 of Ferreccio et al. (2000) are based on average arsenic exposure levels during the high exposure period (1958-1970) and Table 5 uses exposure levels averaged over the entire study period (1930-1994), the reported odds ratios are higher in Table 5 than in Table 6.

Table 5-3 shows the risk ratios associated with exposure to 50 μg/L for the two studies discussed above (Ferreccio et al. 2000; Chiou et al. 2001) and corresponding ED_{01} estimates. Results are also included for the southwestern Taiwanese data previously reported by Chen et al. (1985) and reanalyzed by Morales et al. (2000) but now recalculated using the BEIR IV lifetime-risk formula. The ranges reported along with each risk ratio generally correspond to 95% CIs obtained from fitted models except where risk ratios were taken from published papers, in which case, they refer to feasible ranges obtained by smoothing or extrapolating the published results. For example, in the case of the Ferreccio et al. (2000) data, the subcommittee applied linear regression to the odds ratios reported in the paper and computed confidence limits based on the fitted regression line. It should be emphasized that such calculations are somewhat imprecise as a result and should mainly be used for qualitative comparisons and for providing information on the range of possible ED_{01}s.

Statistical Analyses and Dose-Response Modeling

In addition to the issue of study selection for an arsenic risk assessment, questions have arisen regarding statistical modeling and sources of uncertainty that might affect the ED calculation in EPA's risk assessment. Those issues are discussed below.

Model Choice

Use of a Comparison Population

The 1999 NRC report recommended that calculation of EDs be based on a Poisson regression model or SMR approach applied to the southwestern Taiwanese data (Chen et al. 1985, 1992; Wu et al. 1989). A variety of feasible model formulations were presented (see 1999 report for dose curves), and it was recommended that those models be considered and compared in some type of sensitivity analysis. As discussed earlier, the data comprised age- and gender-specific cancer mortality data from 42 villages in the region of south-western Taiwan where arsenic is endemic. The 1999 NRC report discussed two possible approaches to the dose-response analysis: (1) analyzing only the 42 villages and using the variation in cancer rates from village-to-village to determine the nature of the estimated dose-response relationship (using an internal comparison group); and (2) incorporating data from an external comparison population, for example, nationwide data. The latter approach is classically used in the analysis of cohort data (Breslow and Day 1988) and has the advantage of providing a much more accurate estimate of the baseline cancer rates. Use of an unexposed external comparison population also minimizes the impact of exposure misclassification in the low-dose range within the study population. A potential disadvantage, however, of using an external comparison group is that the analysis can be biased if the study population differs from the comparison population in important ways. Such issues have been discussed extensively in the context of occupational cohort studies, in which, for example, there is often reason to believe that a population of healthy workers might differ from the general population. In the case of the 1999 report, which used a linear dose multiplicative Poisson model, it turned out that ED calculations were very similar, regardless of whether a comparison population was used. Subsequent to the 1999 NRC report, however, Morales et al. (2000) expanded the modeling options beyond those considered by the previous Subcommittee on Arsenic in Drinking Water (NRC 1999) and found considerable model sensitivity. In addition to including exposure as a linear term in the model, Morales et al. (2000) also considered log and square-root transformations and fit additive versions of the dose-response model in addition to the multiplicative models discussed in the 1999 report. As indicated in Table 8 of Morales et al. (2000), ED_{01} estimates from the best-fitting models ranged from 250 to 400 µg/L for male and female lung and bladder cancer, based on models without an external comparison group. The 10- to 150-μg/L range was based on models that used the entire Taiwanese population as a comparison group. In general, estimated ED_{01}s tended to be lower for models that included an external comparison population, primarily because the lung and bladder cancer rates in the comparison populations were much lower than the rates seen even in the study villages with low exposures. Consequently, the fitted dose-response models tended to be steeper (Morales et al. 2000). Furthermore, when comparison populations were included, models based on a log or square-root transformation of dose tended to fit better statistically, since they allowed the dose-response curve to take on a supralinear shape involving a steep initial slope, followed by a flattening out (see Figures 5-1a-c).

As discussed earlier, EPA based its ED on the linear multiplicative model without a comparison population. The SAB recommended this choice primarily because the results based on the no-comparison-group models were more stable (i.e., less sensitive to model choice), and the comparison-group models resulted in extremely low ED estimates, which did not seem heuristically reasonable. Further justification for the choice was provided by the argument that the poor rural areas represented in the study population would not be similar to the comparison populations, which were weighted heavily toward the more prosperous urban areas in Taiwan.

This subcommittee carefully considered the issues surrounding the use of a comparison population and ultimately concluded that a comparison population should be used. Rationale for this conclusion included the following:

- It has been argued that the area of southwestern Taiwan where arsenic is endemic is different from the whole of Taiwan in important ways (other than arsenic in the drinking water) that could affect cancer death rates; hence, national rates were inappropriate to use as a standard referent when calculating SMRs. That argument is not supported by a recent paper by Tsai et al. (1999). As discussed in Chapter 2, those authors demonstrate that SMRs for the area where arsenic is endemic based on regional population rates are similar in magnitude to SMRs based on rates for the national population. (The vast majority of individuals in the southwestern Taiwan referent region are not exposed to increased arsenic concentrations in drinking water.) Therefore, the use of an external comparison population is unlikely to introduce significant confounding.
- Available dose-response data cannot distinguish between linear and nonlinear models.

—Indeed, one of the reasons cited for not using models that incorporated a comparison population is that they tend to produce supralinear dose-response curves, when it has been argued that the mechanistic data point to a sublinear model. As discussed in Chapter 3, however, the mechanistic evidence for a sublinear dose-response relationship do not necessarily apply at the population level. Furthermore, the fact remains that there is some empirical evidence suggesting that a supralinear model might indeed hold. For example, statistical goodness-of-fit criteria applied to both the southwestern Taiwanese data (Chen et al. 1985, 1988, 1992; Wu et al. 1989) and the recent data from northeastern Taiwan (Chiou et al. 2001) support models based on a log transformation of dose, leading to supralinear models. The data from other published papers point to similar patterns. The study by Ferreccio et al. (2000) also yielded some evidence of a supralinear dose-response relationship. A supralinear dose-response relationship is plausible; for example, it could result from a subpopulation that is susceptible to low doses of arsenic.

—Another possible reason for an apparent supralinear curve is dose misclassification. The exposure to high concentrations of arsenic in food (i.e., on the order of 50 µg/day) would result in a substantial increase in the dose of arsenic received by people in villages with low concentrations of arsenic in drinking water. The impact of the added arsenic would be relatively much greater for individuals with low concentrations of arsenic in drinking water than for individuals with high concentrations of arsenic in drinking water. If the true total dose of arsenic in food and water were plotted against the responses, the curve might no longer be supralinear.

—Dose misclassification might also result from the movement of people among the different villages. For example, even though a person might live in one particular village, it is logical to think that he or she might visit neighboring villages on occasion and drink the water there or eat food grown in other villages. The 1999 NRC report also discusses the fact that some of the villages had multiple wells with high variation in measured arsenic concentrations. If the villages assigned low exposure status actually contained a significant number of individuals with higher exposure, such misclassification bias would result in a substantial underestimation of the slope of the dose-response relationship in models limited to the study villages. Results based on models that include an external comparison population are likely to be less affected by such measurement error, because the model fit is anchored by a large control group.

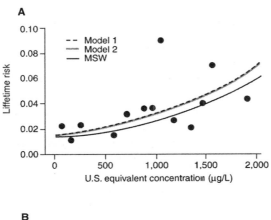

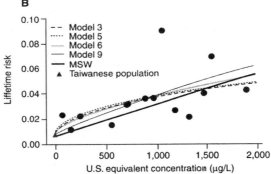

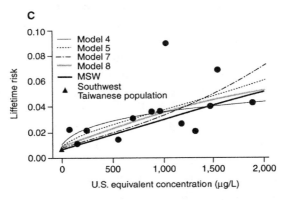

Figure 5-1 Estimated lifetime death risk over background rates in Taiwan for male bladder cancer (A) without comparison population, (B) with Taiwan-wide comparison population, and (C) with southwestern Taiwan comparison population. Source: Morales et al. (2000).

Functional Form of Dose-Response Relationship

As discussed in the previous section, when an external comparison population was included, the ED estimates from the analyses of the southwestern Taiwanese data used by EPA in setting its proposed standard varied depending on the functional form of the dose-response model being fitted. According to a statistical measure of model fit, the Akaike information criteria (AIC) (see Hastie and Tibshirani (1990) for discussion of model-fit criteria), statistically superior fits were produced using models that included a log or square-root transformation of dose (Morales st al. 2000), despite the fact that those models are not as biologically plausible as others. Note that for AIC values a lower number indicates a better fit. Table 5-4 shows AIC values for some of the models shown in Morales et al. (2000) and other models, including one with a quadratic term in dose and one with an interaction between dose and age. Models based on both a multiplicative relative risk and an additive relative risk are presented. The quadratic dose-response model, which was not investigated by Morales et al. (2000), also resulted in a relatively good fit. In addition to the AIC values, Table 5-4 includes a column of Bayesian posterior model probabilities (PMPs) (see Carlin and Louis 1996 for discussion). PMPs are computed by starting with an initial assumption that all models under consideration are equally likely and then computing the conditional probability that each one is true, given the observed data. The effect of these posterior probabilities is to downweight models that do not fit well and emphasize models that do. As can be seen Table 5-4, the PMPs tend to correspond (inversely) reasonably well with the AIC values (i.e., the PMPs are high when AICs are low, and visa versa) but have the important advantage of being easier to interpret. Looking at the PMP values suggests that the additive models, using either a linear or log dose, or the multiplicative model with quadratic dose best fit the data. Only the results for male lung cancer are shown in Table 5-4. Similar patterns emerged for female lung cancer, and male and female bladder cancer. In light of considerations of biological plausibility and consistency with other recommended approaches to quantitative dose-response assessment (Smith et al. 1992), the subcommittee recommends the use of the additive Poisson model with dose entered as a linear term.

In a recent doctoral dissertation, Morales (2001) used the technique of Bayesian model averaging (Carlin and Louis 1996) to compute an ED_{01} that combines the results from a set of models, such as those listed in Table 5-3. The technique can be useful in situations, such as the one for the arsenic data, where several different models are feasible. Another advantage of this tech-

TABLE 5-4   Akaike Information Criteria Values and Bayesian Posterior Model Probabilities[a]

Relative Risk Model	Model Description	AIC Value	PMP Value
$1 + \beta\, d$	Additive, linear dose	425.8	0.152
$\exp(\beta_0 + \beta_1\, d)$	Multiplicative, linear dose	435.9	0.000
$\exp(\beta_0 + \beta_1\, d + \beta_2\, d^\wedge 2)$	Multiplicative, quadratic dose	425.1	0.432
$\exp(\beta_0 + \beta_1\, d + \beta_2\, d*t)$	Multiplicative, interaction between age and dose	437.6	0.000
$1 + \beta\, \log(d)$	Additive, log dose	424.3	0.413
$\exp(\beta_0 + \beta_1\, \log(d))$	Multiplicative, log dose	431.1	0.003
$\exp(\beta_0 + \beta_1\, \log(d) + \beta_2\, \log(d)*t)$	Multiplicative, interaction between age and log dose	432.7	0.000

[a] Values are based on male lung cancer data from southwestern Taiwan (Chen et al. 1985, 1992; Wu et al. 1989), with southwestern Taiwan serving as the comparison population. All models include a quadratic effect of age. Models are distinguished by their assumed relative risk associated with an exposure level, d.

Abbreviations: AIC, Akaike information criteria; BPM, Bayesian posterior model probabilities.

nique is that it appropriately quantifies the variability due to model uncertainty. Because it is a non-standard statistical method that has not been extensively evaluated in the context of environmental risk assessment, the subcommittee does not recommend it as the basis of the arsenic risk assessment.

Impact of Dietary Intake of Arsenic

Some attention has been focused on the issue of differences between the United States and Taiwan with respect to dietary intake of arsenic. As discussed earlier, it has been suggested that effective exposures in Taiwan are higher than represented simply by arsenic concentrations in well water, because (1) the Taiwanese people are exposed to additional arsenic through water used for cooking owing to the predominance of rice and dried sweet potatoes in the diet; and (2) people in the region of Taiwan where arsenic is endemic are exposed to a high general background level of inorganic arsenic in their food (50 μg/day in Taiwan compared with 10 μg/day in the United States) (NRC 1999), although there is substantial uncertainty and variability in both of those estimates.

EPA adjusted its lower-bound risk estimates to account for the extra arsenic, adding 1 L of water to drinking-water consumption to account for water used in cooking and multiplying the risk estimates by the fraction of arsenic contributed by drinking water. The subcommittee believes that the method used by EPA to account for the arsenic present in cooking water is valid and easily accomplished. However, the source of the data underlying EPA's decision to add 1 L of cooking water is not documented. EPA assumed that Taiwanese men and women ate one cup (0.23 L) of dry rice and 2 pounds (0.9 kg) of sweet potatoes each day, although it is not clear how those intake data were estimated by EPA. EPA estimated that the amount of arsenic present in food (before cooking) is higher in Taiwan than the United States. The subcommittee found that there is little evidence to support the estimates used. The subcommittee questions EPA's approach to adjusting for background levels of arsenic in food. A preferable approach would involve refitting the dose-response models with the appropriate background amounts added to individuals living in the region where arsenic is endemic.

The subcommittee addressed the issue of background arsenic in food by adding a constant concentration of arsenic to the exposure rates for all individuals in the study villages. The assumed background rate in food was 30 μg/day (corresponding to 0.6 μg/kg/day, assuming a 50-kg weight for a Taiwanese person). The impact on the ED estimates by adding this amount was relatively small (approximately a 1% increase in ED estimates).

Impact of Variability in Drinking-Water Intake

The first three rows of Table 5-5 show the impact of varying the assumed ratio between mean water-consumption rates in Taiwan and the United States. ED_{01}s and LED_{01}s were calculated using a multiplicative model, because it is computationally easier to implement; similar results would be expected using an additive model. As expected, the assumed ratio has a fairly dramatic impact on estimated ED_{01}s. The subcommittee considered the impact of variation in individual drinking-water intake on risk estimates based on the Taiwanese data used by EPA in arriving at their regulatory level of 10 µg/L. The discussion of this topic has centered around two main issues:

—The Taiwanese study used by EPA was ecological in nature, meaning that exposure was measured at the village level rather than the individual level. Differences among individuals in the amount of water they drink results in interindividual variability in exposure concentrations.

—A typical individual in the United States is likely to have a lower drinking-water intake (per unit of body weight) than a typical individual in Taiwan. It should be noted, however, that the assumption of major differences in water consumption between the Taiwanese study population and the U.S. population has been questioned (Mushak and Crocetti 1995).

To address those issues, a hierarchical model that formulated the true relative-risk model in terms of micrograms of arsenic per unit of body weight was fitted to the data. Drinking-water intake rates for the United States were based on data reported by EPA (2000d). Intake rates were also assumed to follow a gamma distribution with a mean 21 mL/kg/day and a standard deviation of 15 mL/kg/day, which compares well with the actual data shown in Figure 4-1. Distributions of intake rates for Taiwan were also assumed to follow a gamma distribution, but with the mean multiplied either by 1 (implying no difference), 2.2 (the factor used in EPA 1988), or 3 (allowing implicitly for additional arsenic exposure through food and cooking water, as discussed above and used by EPA in its analysis). The models were fit using a Bayesian formulation (Carlin and Louis 1996) in the statistical package BUGS (Spiegelhalter et al. 1996) and a specialized program written for the subcommittee in Fortran. Table 5-5 shows the impact of drinking-water variability on ED_{01} estimates and standard errors. As can be seen in Table 5-5, estimates of the ED_{01}s tend to increase when individual variability in both Taiwan and the United States are taken into account. On the other hand, the lower limits on the ED_{01}s decrease. Therefore, as expected, the spread between the lower limits and the ED_{01}s increases when adjustments are made.

Effects of Exposure-Measurement Error

Measurement errors in assigning village-specific exposures have been cited as another source of concern for analyses based on the Taiwanese data. As indicated in the 1999 NRC report, some of the villages had multiple wells with

TABLE 5-5 Effect of Variability in Water Consumption on 1% Effective Dose Estimates Based on Male Lung Cancer in Southwestern Taiwan[a]

Adjusted for Individual Variability	Taiwanese/U.S. Mean Drinking-Water Rates	ED_{01} (μg/L)	LED_{01} (μg/L)
No	1	65	57
	2.2	145	129
	3	195	173
Yes	1	117	41
	2.2	191	66
	3	246	85

[a] Multiplicative model with exposure entered as a linear term and age as a quadratic. A similar finding would be expected if the calculations were conducted using an additive model. The southwestern region of Taiwan is used as comparison population. BIER IV formula is used for ED_{01} calculations.

Abbreviations: ED_{01}, 1% effective dose; LED_{01}, lower 95% confidence limit on the 1% effective dose.

wide ranges of arsenic concentrations (see Table A10-1, NRC 1999), yet all individuals living in that village would be assigned the median level as their exposure concentration. The statistical theory of measurement error (see Carroll et al. 1995) would predict a bias in estimated EDs under such circumstances. To assess the degree of such bias, a small sensitivity analysis, fitting a range of feasible measurement-error models, was conducted. In the analysis, the concentration observed for the particular village is generated from a gamma distribution whose mean is the unknown true median concentration for that village and whose variance is defined as σ^2. As discussed by Carroll et al. (1995), fitting a measurement-error model also requires specification of the distribution of true values; it was assumed that the true village-specific medians were uniformly distributed between 0 and 1,200 μg/L. Models were fit in the package BUGS (Spiegelhalter et al. 1996). Table 5-6 shows the results of this analysis based on a multiplicative model with a linear dose effect, using the southwestern region of Taiwan as an external comparison population. Although the subcommittee has recommended the use of the additive model, it has used the multiplicative model for the purpose of this exercise, because it is easier to implement computationally. As can be seen in the table, the estimated ED_{01} is attenuated as the magnitude of measurement error increases. The attenuation is modest and very small relative to the variability associated with model uncertainty. Furthermore, the subcommittee is recommending the use of an external comparison population, and under those circumstances, the impact of the measurement error would be expected to be much smaller (analyses not shown). The analysis reported here is based on strong assumptions and should not be overinterpreted as an actual assessment of the measurement error. The analysis is presented solely to assess the possible impact of measurement errors. To address the issue in depth would require more extensive analysis and is beyond the scope of this chapter.

Mortality Versus Incidence

The ED_{01} calculations reported by NRC (1999) and Morales et al. (2000) based on the data from southwestern Taiwan referred to lifetime cancer mortality (bladder or lung). Some assumptions were needed to compare these results with those based on the bladder cancer data from the study of Chiou et al. (2001) in northeastern Taiwan and the data on lung cancer from the study by Ferreccio et al. (2000) in Chile, since both studies report incidence data. In Table 5-3, the subcommittee reports ED_{01}s for cancer mortality based on the results from the Ferreccio et al. (2000) study and the Chiou et al. (2001) study. It was assumed for those calculations that the same relative risks would apply for cancer incidence and cancer mortality. For the purpose of discussing risks in the U.S. population more generally, however, it is risk of disease incidence, not mortality, that is the end point of interest. As discussed earlier, EPA converted risks calculated from Taiwanese mortality data to cancer incidence by assuming an 80% mortality for bladder cancer and a 100% mortality for lung cancer (see Table 5-2). Even though that assumption is probably appropriate for lung cancer, the subcommittee is concerned that a calculation based on the 80% mortality for bladder cancer in Taiwan might not be appropriate for the U.S. population. According to the SEER (2001) cancer registry data, the U.S. lifetime bladder cancer incidence is approximately five times that of lifetime bladder mortality, indicating that the U.S. case mortality is close to 20%.

Calculated Risk Estimates

In this section, the subcommittee reports risk estimates that are based on incidence data taken directly from the SEER (2001) registry. Use of the BEIR

TABLE 5-6 1% Effective Dose and Lower 95% Confidence Limit on 1% Effective Dose Computed Under Varying Degrees of Assumed Measurement Error[a]

Measurement Error, S.D. (σ^2), µg/L	ED_{01} (µg/L)	LED_{01} (µg/L)
0 (0)	145	129
10 (100)	145	129
32 (1,000)	144	127
45 (2,000)	143	127
55 (3,000)	142	125

[a] On village-specific median arsenic levels. Calculations presented for male lung cancer data from the southwestern Taiwanese study using the Poisson multiplicative linear model. The ratio of the Taiwanese-to-U.S. mean drinking-water rate was assumed to be 2.2.
Abbreviations: ED_{01}, 1% effective dose; LED_{01}, lower 95% confidence limit on the 1% effective dose; S.D., standard deviation; σ^2, variance.

IV formula (or its simple approximation—see earlier discussion) easily facilitates this approach. The lifetime cancer incidence assumed for the calculations reported here are 7.85% and 5.75% for male and female lung cancer, respectively, and 3.42% and 1.13% for male and female bladder cancer, respectively. It is useful to note again for comparison purposes that the corresponding lifetime death rates are 7.62% and 4.85% for lifetime lung cancer mortality (male and female, respectively), and 0.73% and 0.3% for lifetime bladder cancer mortality (male and female, respectively) (SEER 2001).

Tables 5-7, 5-8, and 5-9 present estimated lifetime excess risks for lung and bladder cancer incidence for populations exposed to arsenic at 3, 5, 10, and 20 µg/L of drinking water. The risk estimates in Tables 5-7 and 5-8 are for lung and bladder cancer, respectively, and were calculated for the U.S. population using the lifetime background incidence for these diseases noted above. Risk estimates in Table 5-9 are adjusted by the difference in background incidence for bladder and lung cancer between the United States and Taiwan. Although lifetime incidences for these cancers in Taiwan are not readily available, age-adjusted incidences for bladder and lung cancer have been reported for the period 1993-1997 (You et al. 2001). Assuming that life expectancy is not dramatically different between Taiwan and the United States during this period, the ratio of age-adjusted incidence reported for Taiwan can be compared with that reported for the United States during a similar period from the SEER (2001) data. For the same period for which data are available for Taiwan (1993-1997), the age-adjusted incidence of lung cancer in the United States was 58.6 (males) and 34.0 (females) per 100,000. The background incidence of lung cancer in the United States, relative to Taiwan, is about 3 times greater for both males and females. Likewise, the background incidence for bladder cancer in the United States was 23.3 (males) and 5.4 (females) per 100,000. Thus, the background incidence for bladder cancer in the United States, relative to Taiwan, was approximately 3.4-fold higher for males and 1.6-fold higher for females. Those factors have been used to calculate estimated risks using the background incidence for bladder and lung cancer of Taiwan (Table 5-9) rather than that of the United States (Tables 5-7 and 5-8). The net effect of using U.S. background rates results in an increase in the risk estimates for arsenic-related cancers in the U.S. populations relative to what is obtained if the background rate of Taiwan is used, projecting a greater risk per unit dose of arsenic in the United States. Although not common practice for some researchers outside the field of epidemiology, the use of relative risks to infer risks across different populations is a widely accepted practice in epidemiology and has considerable scientific support (Breslow and Day 1988). It should be noted that the previous NRC (1999) report and the EPA's risk assessment (EPA 2001) did not use the higher U.S. background rates for bladder and lung cancer in their final risk estimates, although the previous NRC report discusses an approach using U.S. background rates (NRC 1999), and the previous subcommittee considered it a valid option.

The maximum-likelihood estimates (MLE) in Table 5-7 and the lower and upper confidence limits on those estimates are presented. It is important to note that the confidence limits, which for the Taiwanese data are all less than 12% above or below the MLE, reflect only statistical variability, not uncertainty resulting from the issues discussed earlier in this chapter and in Chapter 4, including interindividual variability. For that reason, the confidence limits are presented in Table 5-7 as examples, but the individual limits are not presented in other tables. The choice of which point on the distribution to use for regulatory purposes, be it the MLE or the upper or lower confidence limit, is a policy choice.

All risk estimates presented in Tables 5-7, 5-8, and 5-9 are calculated using risk ratios determined by the additive Poisson model with dose entered as a linear term and using the BEIR IV formula. For the Taiwanese data set, the mortality data from all of southwestern Taiwan was used in the model to represent an unexposed external comparison population. It should be noted

TABLE 5-7 Theoretical Excess Lifetime Risk Estimates (Incidence per 10,000 people) of Lung Cancer for U.S. Populations Exposed at Various Concentrations of Arsenic in Drinking Water Using Chilean and Taiwanese Data[a]

| Arsenic Concentration (µg/L) | Maximum-Likelihood Estimate (95% Lower Limit, 95% Upper Limit) | | | |
| | Chilean Data[b] | | Taiwanese Data[c] | |
	Females	Males	Females	Males
3	14 (11, 18)	20 (14, 25)	5.4 (5.1, 5.7)	4.0 (3.9, 4.2)
5	24 (18, 29)	33 (24, 42)	8.9 (8.5, 9.4)	6.8 (6.5, 7.0)
10	48 (36, 59)	67 (48, 83)	18 (17, 19)	14 (13, 14)
20	95 (71, 120)	130 (95, 170)	36 (34, 38)	27 (26, 28)

[a] These risks are estimated using assumptions considered to be reasonable by the subcommittee; it is possible to get higher and lower estimates using other assumptions. Confidence limits presented reflect statistical variability only, reflecting primarily the sample size. As such, they are not indicative of the true uncertainty associated with the estimates. Risk estimates are rounded to two significant figures.

[b] Based on the data from Chile using the average arsenic concentration from the peak years of exposure (1958 to 1970) (Ferreccio et al. 2000), assuming that the typical U.S. and Chilean resident weighs 70 kg, that drinking rates in both countries are the same, and hence that the Chilean exposure per kilogram of body weight is 1.0 times the U.S. exposure, calculated using an additive Poisson model with linear dose. Risks were also estimated using the same assumptions but based on the average arsenic concentrations from 1930 to 1994. The risk estimates (per 10,000) and corresponding upper and lower confidence limits in females exposed at 3, 5, 10, and 20 µg/L are 50 (33, 75), 83 (56, 130), 170 (110, 250), and 330 (220, 500) respectively. The corresponding estimates for males are 75 (43, 100), 130 (71, 170), 250 (140, 330), and 500 (290, 670) respectively.

[c] Estimates were calculated using data from individuals in the arsenic-endemic region of southwestern Taiwan (Chen et al. 1985, 1992; Wu et al. 1989), and data from an external comparison group from the overall southwestern Taiwan area. It was assumed that the typical U.S. resident weighs 70 kg, compared with 50 kg for the typical Taiwanese, and that the typical Taiwanese drinks just over 2 L of water per day, compared with 1 L per day in the United States; thus, the Taiwanese exposure per kilogram of body weight is 3 times the U.S. exposure, calculated using an additive Poisson model with linear dose.

that the vast majority of individuals in all of southwestern Taiwan are not exposed to increased arsenic concentrations in drinking water.

To permit comparisons of cancer risks for Taiwan with those for Chile and United States, the subcommittee has calculated risk estimates using different studies and assumptions. Thus, a range of possible risks from arsenic in drink-

TABLE 5-8 Theoretical Maximum-Likelihood Estimates of Excess Lifetime Risk (Incidence per 10,000 People) of Bladder Cancer for U.S. Populations Exposed at Various Concentrations of Arsenic in Drinking Water Using Different Ratios for Taiwanese-to-U.S. Drinking-Water Ingestion on a Per-Body-Weight Basis[a]

Arsenic Concentration (µg/L)	Taiwanese to U.S. Drinking Water Ratio = 1.4[b]		Taiwanese to U.S. Drinking Water Ratio = 3[c]	
	Females	Males	Females	Males
3	7.7	15	3.6	6.8
5	13	25	6.0	11
10	26	50	12	23
20	51	100	24	45

[a] These risks are estimated using assumptions considered to be reasonable by the subcommittee; it is possible to get higher and lower estimates using other assumptions. Estimates were calculated using data from individuals in the arsenic-endemic region of southwestern Taiwan (Chen et al. 1985, 1992; Wu et al. 1989), and data from an external comparison group from the overall southwestern Taiwan area. Risk estimates are rounded to two significant figures. All 95% confidence limits are less than +/- 12% of the maximum-likelihood estimate and are not presented. Those confidence limits reflect statistical variability only, reflecting primarily the sample size. As such, they are not indicative of the true uncertainty associated with the estimates.
[b] Risks were estimated assuming that the typical U.S. resident weighs 70 kg, compared with 50 kg for the typical Taiwanese, and that the typical Taiwanese drinks just over 1 L of water per day, the same as the 1 L per day in the United States; thus, the Taiwanese exposure per kilogram of body weight is 1.4 times the U.S. exposure, calculated using an additive Poisson model with linear dose.
[c] Risks were estimated assuming that the typical U.S. resident weighs 70 kg, compared with 50 kg for the typical Taiwanese, and that the typical Taiwanese drinks just over 2 L of water per day, compared with 1 L per day in the United States; thus, the Taiwanese exposure per kilogram of body weight is 3 times the U.S. exposure, calculated using an additive Poisson model with linear dose.

ing water are presented. Those values, however, should not be considered bounds on the possible risk estimates, because other assumptions could be made that would result in higher or lower values. When estimating the risks to the U.S. population, assumptions must be made about body weights and water consumption in both the United States and the Taiwanese populations. Then, when comparing cancer risk estimates, it is important to be aware of how those assumptions affect the estimates. For example, the higher the ratio of water ingestion in Taiwan relative to the United States in terms of liters per body weight per day, the smaller the U.S. cancer risk estimate will be. The assumptions that the subcommittee used to calculate the various estimates are discussed below and presented in the footnotes of the tables.

Table 5-7 presents estimates for the excess lifetime risk (incidence) of lung cancer in females and males in the U.S. population from exposure to arsenic at 3, 5, 10, and 20 µg/L of drinking water. The estimates based on the Chilean data (Ferreccio et al. 2000) were calculated assuming that the typical U.S. and Chilean resident both weigh 70 kg and that the drinking-water ingestion rates in both countries are the same—that is, that exposures in Chile on a per-body-weight basis, given the same concentration of arsenic in the drinking water, are equal to those in the United States (i.e, a factor of 1). The estimates based on the southwestern Taiwanese data set (Chen et al. 1985, 1992; Wu et al. 1989), however, were calculated assuming that the typical U.S. resident weighs 70 kg and drinks 1 L of water per day, and the typical Taiwanese resident weighs 50 kg and drinks just over 2 L of water per day. Therefore, given the same concentration of arsenic in drinking water, exposures in Taiwan on a per-body-weight basis would be three times that of a U.S. resident. The data in Table 5-7 utilize the background rate of lung cancer in the United States. To illustrate the importance of the background rate, Table 5-9 shows the same risk projections using the Taiwanese data set and the background

TABLE 5-9 Theoretical Maximum-Likelihood Estimates of Excess Lifetime Risk (Incidence per 10,000 people) of Lung and Bladder Cancer for Populations Exposed at Various Concentrations of Arsenic in Drinking Water, Using the Background Cancer Incidence Rate for Taiwan[a]

Arsenic Concentration (µg/L)	Bladder Cancer		Lung Cancer	
	Females	Males	Females	Males
3	2.3	2.0	1.8	1.7
5	3.8	3.2	3.0	3.0
10	7.5	6.8	6.2	6.1
20	15	13	12	12

[a] These risks are estimated using the assumptions noted in footnotes (a) and (c) of Table 5-8, and assuming a Taiwanese to U.S. drinking water ratio of 3. The estimated background incidence rates for bladder cancer in Taiwan (derived from the subcommittee's cancer risk estimates presented in Tables 5-7 and 5-8 and adjusted by the ratio of Taiwanese incidence data (You et al.2001) to U.S. incidence data (Ferlay et al. 2001)) are 6.9 (males) and 3.4 (females) per 100,000 and for lung cancer are 25.8 (males) and 11.9 (females) per 100,000.

incidence rates for lung cancer in Taiwan, which are approximately 2- and 3-fold lower than those in the United States for males and females, respectively.

For lung cancer, the MLEs of risk for males based on the Chilean data (Ferriccio et al. 2000) range from 20 per 10,000 at 3 μg/L to 130 per 10,000 at 20 μg/L using the average arsenic concentration during the peak exposure period of 1958 to 1970. The corresponding risk estimates in females are 14 per 10,000 to 95 per 10,000. Using the average arsenic concentration from 1930 to 1994, the risk estimates range from 75 to 500 per 10,000 males and from 15 to 330 per 10,000 females. Using the southwestern Taiwanese data, the risk estimates for arsenic at 3 μg/L of drinking water range from 1.7 to 4.0 per 10,000 for males and from 1.8 to 5.4 per 10,000 for females, depending on which assumptions are used (Tables 5-8 and 5-9).

Different studies have estimated the risks of lung cancer following exposure to arsenic, and it is possible and useful to compare the risk estimates generated in the different analyses at a given arsenic concentration. Because the case mortality for lung cancer is close to 100% (SEER 2001), lung cancer incidence approximates lung cancer mortality, and the risk estimates for lung cancer mortality can be compared with risk estimates for lung cancer incidence. As can be seen in Table 5-7, the subcommittee's analysis of the southwestern Taiwanese data yields an estimate for lifetime lung cancer incidence in the United States (using the U.S. background rate) at an arsenic concentration in drinking water of 10 μg/L of approximately 14 per 10,000 and 18 per 10,000 in males and females, respectively. It is noteworthy that nearly 10 years ago, Smith et al. (1992) published a risk assessment based on the same ecological southwestern Taiwanese data analyzed by the subcommittee. In that risk assessment, lifetime lung cancer mortality risks at 10 μg/L were estimated to be 11 per 10,000 and 17 per 10,000 for males and females, respectively. Therefore, lung cancer risk estimates generated by this subcommittee and those published by Smith et al. (1992) are very consistent. The lung cancer risk estimates derived from the Taiwanese data in Table 5-7 can be compared with those derived from the data in the recent case-control study in northern Chile by Ferreccio et al. (2000). When the average arsenic concentration in northern Chile from 1930 to 1994 is used as the dose-metric, the risk estimates for lung cancer incidence in the United States calculated by the subcommittee for an arsenic concentration of 10 μg/L are 250 per 10,000 males and 170 per 10,000 females. Those estimates are approximately an order of magnitude higher than the estimates the subcommittee derived from the Taiwanese data. However, when the dose metric selected for the Chilean data is the peak years of arsenic exposure from 1958 to 1970, the corresponding U.S. lung cancer risk estimates are 67 per 10,000 males and 48 per 10,000 females. Those estimates are approximately 3 to 4 times higher than the subcommittee's estimates derived from the Taiwanese data. For comparative purposes, the subcommittee also derived cancer risk estimates at 10 µg/L using the relative risks (see Table 2-1) for lung cancer associated with the peak period of arsenic exposure in northern Chile in the ecological study by Smith et al. (1998). If the same formula is applied to those relative risks, as was used to estimate the subcommittee's other cancer estimates, the U.S. lung cancer estimates at 10 µg/L are 38 per 10,000 and 21 per 10,000 in males and females, respectively.

Overall the peak period exposure data in northern Chile and the data from southwestern Taiwan yield coherent lifetime excess risk estimates ranging from 1.4 to 6.7 per 1,000 for lung cancer in the United States at a drinking-water arsenic concentration of 10 µg/L. The finding that risk estimates derived from studies of individuals exposed to arsenic in Chile are similar to those estimated from Taiwanese data provides confidence in the validity of the risk estimates.

Tables 5-8 and 5-9 present estimates for the excess lifetime risk (incidence) for bladder cancer in females and males in the U.S. population from exposure to arsenic at 3, 5, 10, and 20 µg/L of drinking water based on the southwestern Taiwanese study (Chen et al. 1985, 1992; Wu et al. 1989), using either the U.S. background rate for bladder cancer (Table 5-8) or the Taiwanese background rate (Table 5-9). In one water-intake scenario, the subcommittee assumed that the typical U.S. resident weighs 70 kg and drinks 1 L of water per day, and the typical Taiwanese resident weighs 50 kg and also drinks 1 L of water per day. Therefore, given the same concentration of arsenic in drinking water, exposures in Taiwan on a per-body-weight basis are 1.4 times that of a U.S. resident. In a second water-intake scenario, calculated using the same data set, it was assumed that the typical U.S. resident weighs 70 kg and drinks 1 L of water per day, and the typical Taiwanese resident weighs 50 kg and drinks just over 2.2 L of water per day. Therefore, as indicated in the lung cancer estimates earlier at the same concentration of arsenic in drinking water, exposures in Taiwan on a per-body-weight basis are 3 times that of a U.S. resident.

For bladder cancer, the maximum-likelihood estimate of risk using a ratio of 1.4 for Taiwanese-to-U.S. drinking-water rates on a per-body-weight basis for males range from approximately 15 per 10,000 at an arsenic concentration of 3 µg/L of drinking water to 100 per 10,000 at 20 µg/L. The corresponding estimates in females range from 8 to 51 per 10,000. Using a drinking water ratio of 3.0, the corresponding ranges are approximately 7 to 45 per 10,000 in males and 4 to 24 per 10,000 in females, using the U.S. background rate. Risk estimates are 3-fold lower for males and 2-fold lower for females if the Taiwanese background rate is used (Table 5-9).

Further discussion of the subcommittee's quantitative risk estimates and comparison of them with the results of other analyses are presented in Chapter 6.

SUMMARY AND CONCLUSIONS

- Since EPA issued a pending standard of 10 µg/L, based on lung and bladder cancer data from the southwestern Taiwanese study in 2000, two additional studies have appeared in the literature that are of sufficient size and quality and with adequate quantification of dose to be considered in computing EDs for arsenic in drinking water. One is a study that examined urinary tract cancer, and TCC in particular, in northeastern Taiwan (Chiou et al. 2001); the second is a case-control study of lung cancer in Chile (Ferreccio et al. 2000).

- Although it can be argued that an external comparison group for dose-response analysis of the original Taiwanese data should not be used, the subcommittee believes that such arguments are outweighed by evidence in favor of using a comparison population. A recent paper by Tsai et al. (1999) decreases concerns about the potential role of confounding in using either the southwestern Taiwanese population or the entire Taiwanese population as an external comparison group.

- Poisson models provide a flexible and useful framework for analysis of cohort data of the form seen in both the southwestern and northeastern Taiwanese studies. Although dose can be entered into the model in a variety of ways, the additive model with a linear dose effect is consistent with other analyses that have been applied to cancer cohort data.

- The BEIR IV formula provides a useful approach to computing an ED_{01} for the United States based on relative risks obtained from a different population. The BEIR IV formula allows the incorporation of relative risk estimates obtained from case-control studies. The simple formula presented by Smith et al. (1992) is a close approximation to the BEIR IV approach.

- There is insufficient knowledge on the mode of action of arsenic to justify the choice of any specific dose-response model—the subcommittee explored a variety of models and ultimately used the additive model with linear dose effect. Although different results might be obtained from other reasonable model choices, the estimates do not differ by more than an order of magnitude.

• Accounting for individual exposure rate variability causes the uncertainty in the estimate to increase. Therefore, as shown in Table 5-5, the central tendency estimates of the ED_{01} increase when individual variability in drinking-water rates is considered. The corresponding lower bound estimates for each ED_{01} decreases, however, because the variance about the mean becomes larger.

• Assumptions regarding the differences between mean drinking-water intakes in Taiwan and the United States can have substantial impacts on the estimated ED_{01}. There is evidence that drinking-water intakes in Taiwan might be closer to U.S. intakes than previously suggested, and that is an important source of uncertainty. Increasing the assumed drinking-water intakes for Taiwan provides a suitable approach to adjusting for arsenic exposure via cooking water, although the addition of 1 L, especially for women, seems excessive and warrants further investigation.

• Adjusting for background arsenic exposure through food is likely to have only a modest effect on the estimated ED_{01}.

• Measurement error in assigning the village-specific arsenic exposure concentrations is also likely to have only a modest impact on the estimated ED_{01}, compared with the variability associated with model uncertainty. Further exploration of this issue, however, would be useful.

• Although there are insufficient data available for a formal combined analysis, it is helpful to compare the results across studies and also across various feasible assumptions to obtain a sense of the magnitude of likely effects. Table 5-3 presents such an analysis.

• The northeastern Taiwanese study has several strengths, including exposure assessment and data on potential confounders, which could inform the dose-response assessment. At present, however, the follow-up time is insufficient to provide the precision necessary for quantitative dose-response assessment.

• Analysis of the data from the period of peak arsenic exposure in northern Chile and the data from southwestern Taiwan results in similar estimates of lifetime lung cancer incidence in the United States. The consistency of the results adds to the confidence in the validity of the risk estimates.

RECOMMENDATIONS

● Quantitative risk assessment for arsenic in drinking water should consider the results from the following two studies:

—The data set from the southwestern Taiwanese study reported by Chen et al. (1988, 1992;) and Wu et al. (1989), which was considered by the NRC in the 1999 report and by EPA (lung and bladder cancer in males and females).

—The Chilean case-control study (lung cancer) reported by Ferreccio et al. (2000).

● Dose-response analysis of the southwestern Taiwanese data should incorporate an unexposed comparison group; the southwestern Taiwanese region is the recommended comparison group.

● The current mode-of-action data are insufficient to guide the selection of a specific dose-response model. The additive Poisson model with a linear term in dose is a biologically plausible model that provides a satisfactory fit to the epidemiological data and represents a reasonable model choice for use in the arsenic risk assessment.

● Research should be conducted on techniques for integrating the results from many epidemiological studies into a risk assessment.

● Information on remaining uncertainties should be incorporated into future analyses when it is acquired.

REFERENCES

Albores, A., M.E. Cebrian, I. Tellez, and B. Valdez. 1979. Comparative study of chronic hydroarsenicism in two rural communities in the Laguna region of Mexico. [in Spanish]. Bol. Oficina Sanit. Panam. 86(3):196-205.

Breslow, N.E., and N.E. Day. 1988. Statistical Methods in Cancer Research: Vol. 2. The Design and Analysis of Cohort Studies. New York: Oxford University Press.

Carlin, B.P., and T.A. Louis. 1996. Bayes and Empirical Bayes Methods for Data Analysis. New York: Chapman & Hall.

Carroll, R.J., D. Ruppert, and L.A. Stefanski. 1995. Measurement Error in Nonlinear Models. New York: Chapman & Hall.

CDC (Centers for Disease Control and Prevention). 2001. GMWK I Total Deaths for Each Cause by 5-Year Age Groups, United States, 1993, 1994, 1995, 1996, 1997, and 1998. Mortality Tables. DataWarehouse. National Center for Health Statistics, Centers for Disease Control and Prevention, U.S. Department of Health and Human Services, Hyattsville, MD. Online. Available: http://www.cdc.gov/nchs/datawh/statab/unpubd/mortabs/gmwki.htm. [Sept. 21, 2001].

Cebrian, M. 1987. Some Potential Problems in Assessing the Effects of Chronic Arsenic Exposure in North Mexico [preprint extended abstract]. Preprint of paper presented at the 194th National Meeting of the American Chemical Society, Aug. 30-Sept.4, 1987, New Orleans, LA.

Cebrian, M.E., A. Albores, M. Aguilar, and E. Blakely. 1983. Chronic arsenic poisoning in the north of Mexico. Hum. Toxicol. 2(1):121-133.

Chen, C.J., Y.C. Chuang, T.M. Lin, and H.Y. Wu. 1985. Malignant neoplasms among residents of a blackfoot disease-endemic area in Taiwan: high-arsenic artesian well water and cancers. Cancer Res. 45(11 Pt 2):5895-5899.

Chen, C.J., M. Wu, S.S. Lee, J.D. Wang, S.H. Cheng, and H.Y. Wu. 1988. Atherogenicity and carcinogenicity of high-arsenic artesian well water. Multiple risk factors and related malignant neoplasms of blackfoot disease. Arteriosclerosis 8(5):452-460.

Chen, C.J., C.W. Chen, M.M. Wu, and T.L. Kuo. 1992. Cancer potential in liver, lung, bladder and kidney due to ingested inorganic arsenic in drinking water. Br. J. Cancer 66(5):888-892.

Chiou, H.Y., S.T. Chiou, Y.H. Hsu, Y.L. Chou, C.H. Tseng, M.L. Wei, and C.J. Chen. 2001. Incidence of transitional cell carcinoma and arsenic in drinking water: a follow-up study of 8,102 residents in an arseniasis-endemic area in northeastern Taiwan. Am. J. Epidemiol. 153(5):411-418.

Concha, G., B. Nermell, and M. Vahter. 1998. Metabolism of inorganic arsenic in children with chronic high arsenic exposure in northern Argentina. Environ. Health Perspect. 106(6):355-359.

Cuzick, J., S. Evans, M. Gillman, and D.A. Price Evans. 1982. Medicinal arsenic and internal malignancies. Br. J. Cancer 45(6):904-911.

EPA (U.S. Environmental Protection Agency). 1988. Special report on Ingested Inorganic Arsenic: Skin Cancer; Nutritional Essentiality. EPA/625/3-87/013. Risk Assessment Forum, U.S. Environmental Protection Agency, Washington, DC. July 1988.

EPA (U.S. Environmental Protection Agency). 1996. Proposed Guidelines for Carcinogenic Risk Assessment. Notice. Fed. Regist. 61(79):17960.

EPA (U.S. Environmental Protection Agency). 1998. Cost of Lung Cancer Chapter II.V in Cost of Illness Handbook. Office of Pollution Prevention and Toxics, U.S. Environmental Protection Agency. [Online]. Available: http://www.epa.gov/oppt/ coi/. [August 17, 2001].

EPA (U.S. Environmental Protection Agency). 2000a. 40 CFR Parts 141 and 142. National Primary Drinking Water Regulations. Arsenic and Clarifications to Compliance and New Source Contaminants Monitoring. Notice of proposed rulemaking. Fed. Regist. 65(121):38887-38983.

EPA (U.S. Environmental Protection Agency). 2000b. 40 CFR Parts 141 and 142. National Primary Drinking Water Regulations. Arsenic and Clarifications to Compliance and New Source Contaminants Monitoring. Notice of data availability. Fed. Regist. 65(204): 63027-63035.

EPA (U.S. Environmental Protection Agency). 2000c. Arsenic Proposed Drinking Water Regulation: A Science Advisory Board Review of Certain Elements of the Proposal, A Report by the EPA Science Advisory Board. EPA-SAB-DWC-01-001. Science Advisory Board, U.S. Environmental Protection Agency, Washington, DC. December. [Online]. Available: http://www.epa.gov/sab/ fiscal01.htm.

EPA (U.S. Environmental Protection Agency). 2000d. Estimated Per Capita Water Ingestion in the United States: Based on Data Collected by the United States Department of Agriculture's (USDA) 1994-1996 Continuing Survey of Food Intakes by Individuals. EPA-822-00-008. Office of Water, Office of Standards and Technology, U.S. Environmental Protection Agency. April 2000.

EPA (U.S. Environmental Protection Agency). 2001. 40 CFR Parts 9, 141 and 142. National Primary Drinking Water Regulations. Arsenic and Clarifications to Compliance and New Source Contaminants Monitoring. Final Rule. Fed. Regist. 66(14):6975-7066.

Ferlay, J., F. Bray, P. Pisani, and D. M. Parkin. 2001. GLOBOCAN 2000: Cancer Incidence, Mortality and Prevalence Worldwide, Version 1.0. IARC CancerBase, No. 5. IARC Press, Lyons, France. [Online]. Available: http://www.dep.iarc.fr/ globocan/ globocan.htm. [Sept. 20, 2001].

Ferreccio, C., C. Gonzalez, V. Milosavljevic, G. Marshall, A.M. Sancha, and A.H. Smith. 2000. Lung cancer and arsenic concentrations in drinking water in Chile. Epidemiology 11(6):673-679.

Gail, M. 1975. Measuring the benefit of reduced exposure to environmental carcinogens. J. Chronic Dis. 28(3):135-147.

Hastie, T., and R. Tibshirani. 1990. Generalized Additive Models. New York: Chapman and Hall.

Hopenhayn-Rich, C., M.L. Biggs, A. Fuchs, R. Bergoglio, E.E. Tello, H. Nicolli, and A.H. Smith. 1996. Bladder cancer mortality associated with arsenic in drinking water in Argentina. Epidemiology 7(2):117-124.

Hopenhayn-Rich, C., M.L. Biggs, and A.H. Smith. 1998. Lung and kidney cancer mortality associated with arsenic in drinking water in Córdoba, Argentina. Int. J. Epidemiol. 27(4):561-569.

Hopenhayn-Rich, C., S.R. Browning, I. Hertz-Picciotto, C. Ferreccio, C. Peralta, and H. Gibb. 2000. Chronic arsenic exposure and risk of infant mortality in two areas of Chile. Environ. Health Perspect. 108(7):667-673.

IARC (International Agency for Research on Cancer). 1999. Cancer Survival in Developing Countries, R. Sankaranarayanan, R.J. Black, and D.M. Parkin, eds. Pub. No. 145. Lyon: International Agency for Research on Cancer, World Health Organization.

Kurttio, P., E. Pukkala, H. Kahelin, A. Auvinen, and J. Pekkanen. 1999. Arsenic concentrations in well water and risk of bladder and kidney cancer in Finland. Environ. Health Perspect. 107(9):705-710.

Lai, M.S., Y.M. Hsueh, C.J. Chen, M.P. Shyu, S.Y. Chen, T.L. Kuo, M.M. Wu, and T.Y. Tai. 1994. Ingested inorganic arsenic and prevalence of diabetes mellitus. Am. J. Epidemiol. 139(5):484-492.

Lewis, D.R., J.W. Southwick, R. Ouellet-Hellstrom, J. Rench, and R.L. Calderon. 1999. Drinking water arsenic in Utah: a cohort mortality study. Environ. Health Perspect. 107(5):359-365.

Mazumder, D.N., J. Das Gupta, A. Santra, A. Pal, A. Ghose, S. Sarkar, N. Chattopadhaya, and D. Chakraborty. 1997. Non-cancer effects of chronic arsenicosis with special reference to liver damage. Pp. 112-123 in Arsenic Exposure and Health Effects, C.O. Abernathy, R. L. Calderon, and W. Chappell, eds. London: Chapman & Hall.

Morales, K.H. 2001. Statistical Methods for Risk Assessment Based on Epidemiological Data. Ph. D. Thesis Submitted to Harvard School of Public Health, Department of Biostatistics, Boston, Mass. May 2001.

Morales, K.H., L. Ryan, T.L. Kuo, M.M. Wu, and C.J. Chen. 2000. Risk of internal cancers from arsenic in drinking water. Environ. Health Perspect. 108(7):655-661.

Morris, J.S., M. Schmid, S. Newman, P.J. Scheuer, and S. Sherlock. 1974. Arsenic and noncirrhotic portal hypertension. Gastroenterology 66(1):86-94.

Mushak, P., and A.F. Crocetti. 1995. Risk and revisionism in arsenic cancer risk assessment. Environ. Health Perspect. 103(7-8):684-689.

Nevens, F., J. Fevery, W. Van Steenbergen, R. Sciot, V. Desmet, and J. De Groote. 1990. Arsenic and noncirrhotic portal hypertension: a report of eight cases. J. Hepatol. 11(1):80-85.

Neubauer, O. 1947. Arsenical cancer: a review. Br. J. Cancer 1(June):192-251.

NRC (National Research Council). 1983. Risk Assessment in the Federal Government: Managing the Process. Washington, DC: National Academy Press.

NRC (National Research Council). 1988. Health Risks of Radon and Other Internally Deposited Alpha-Emitters: BEIR IV. Washington, DC: National Academy Press.

NRC (National Research Council). 1999. Arsenic in Drinking Water. Washington, DC: National Academy Press.

Rahman, M., and J.O. Axelson. 1995. Diabetes mellitus and arsenic exposure: A second look at case-control data from a Swedish copper smelter. Occup. Environ. Med. 52(11):773-774.

Rahman, M., M. Tondel, S.A. Ahmad, and C. Axelson. 1998. Diabetes mellitus associated with arsenic exposure in Bangladesh. Am. J. Epidemiol. 148(2):198-203.

Roth, F. 1956. Concerning chronic arsenic poisoning of the Moselle wine growers with special emphasis on arsenic carcinomas. Krebsforschung 61:287-319.

SEER (Surveillance, Epidemiology, and End Results). 2001. Surveillance, Epidemiology, and End Results Program Public-Use Data (1973-1998), National Cancer Institute, DCCPS, Surveillance Research Program, Cancer Statistics Branch. [Online]. Available: http://www-seer.ims.nci.nih.gov/Publications/. [August 20, 2001].

Smith, A.H., C. Hopenhayn-Rich, M.N. Bates, H.M. Goeden, I. Hertz-Picciotto, H.M. Duggan, R. Wood, M.J. Kosnett, and M.T. Smith. 1992. Cancer risks from arsenic in drinking water. Environ. Health Perspect. 97:259-267.

Smith, A.H., M. Goycolea, R. Haque, and M.L. Biggs. 1998. Marked increase in bladder and lung cancer mortality in a region of northern Chile due to arsenic in drinking water. Am. J. Epidemiol. 147(7):660-669.

Spiegelhalter, D.J., A. Thomas, N.G. Best, and W.R. Gilks. 1996. BUGS: Bayesian inference Using Gibbs Sampling, Version 0.5. Cambridge: MRC Biostatistics Unit. [Online]. Available: http://www.mrc-bsu.cam.ac.uk/bugs/. [August 20, 2001].

Tsai, S.M., T.N. Wang, and Y.C. Ko. 1999. Mortality for certain diseases in areas with high levels of arsenic in drinking water. Arch. Environ. Health 54(3):186-193.

Tseng, W.P. 1977. Effects and dose-response relationships of skin cancer and blackfoot disease with arsenic. Environ. Health Perspect. 19:109-119.

Tseng, W.P., H.M. Chu, S.W. How, J.M. Fong, C.S. Lin, and S. Yeh. 1968. Prevalence of skin cancer in an endemic area of chronic arsenicism in Taiwan. J. Natl. Cancer Inst. 40(3):453-463.

Tsuda, T., A. Babazono, E. Yamamoto, N. Krumatani, Y. Mino, T. Ogawa, Y. Kishi, and H. Aoyama. 1995. Ingested arsenic and internal cancer: A historical cohort study followed for 33 years. Am. J. Epidemiol. 141(3):198-209.

Wu, M.M., T.L. Kuo, Y.H. Hwang, and C.J. Chen. 1989. Dose-response relation between arsenic concentration in well water and mortality from cancers and vascular diseases. Am. J. Epidemiol. 130(6):1123-1132.

Yeh, S. 1973. Skin cancer in chronic arsenicism. Hum. Pathol. 4(4):469-485.

You, S.L., T.W. Huang, and C.J. Chen. 2001. Cancer registration in Taiwan. Asian Pac. J. Cancer Prev. 2(IACR suppl.):75-78.

Zaldivar, R. 1974. Arsenic contamination of drinking water and foodstuffs causing endemic chronic poisoning. Beitr. Pathol. 151(4):384-400.

6

Hazard Assessment

After completing a hazard identification and dose-response assessment, it is important to place the conclusions of the assessment in the context of similar analyses and in the context of a real-world situation to ensure that the risk estimates are reasonable in view of available information. Therefore, this chapter will summarize the findings of the subcommittee, compare the results of the subcommittee's dose-response assessment with those of the previous NRC (1999) subcommittee and EPA (2001), and, finally, examine whether the estimated risks are plausible when considered in the context of the U.S. population.

FINDINGS OF THE SUBCOMMITTEE

There is increasing evidence that chronic exposure to arsenic in drinking water may be associated with an increased risk of hypertension and diabetes. The existing studies from Taiwan and Bangladesh, discussed in Chapter 2, have observed substantial increases in the risk of these medical conditions at levels of arsenic exposure that are within one to two orders of magnitude of the lower levels of current regulatory concern in the United States. Pending further research that characterizes the dose-response relationship for these end points, the magnitude of possible risk that exists at low levels is nonquantifiable. Nevertheless, because these end points are common causes of morbidity and mortality, even small increases in relative risk at low dose could be of considerable public-health importance. This potential impact should be qualitatively considered in the risk-assessment process.

A sound and sufficient database exists on the carcinogenic effects of arsenic in humans. The main end points for a quantitative risk assessment following exposure to arsenic in drinking water are lung and bladder cancers. The human data from southwestern Taiwan used by EPA in its risk assessment remain the most appropriate for determining quantitative risk estimates. Human data from more recent studies provide additional information for use in the risk assessment. Based on some of these studies, the subcommittee recommends using an external comparison population when analyzing the earlier studies from southwestern Taiwan, rather than comparing high- and low-exposure groups within the exposed population, because of concerns regarding probable exposure misclassification in the low-exposure villages within the data set and because of new data from southwestern Taiwan that suggest that confounding is unlikely. The data on the mode of action of arsenic do not indicate what form of extrapolation should be used below the exposure range of human data. The observed data should be modeled using a biologically plausible model form that best fits the data to determine a 1% effective dose (ED_{01}). The subcommittee used an additive Poisson model with a linear term in dose for the southwestern Taiwan cancer data. The dose-response relationship should be extrapolated linearly from the ED_{01} to zero. Because the human data include exposures to arsenic concentrations relatively close to some U.S. exposures, the distance of extrapolation is very small—less than 1 order of magnitude.

The subcommittee calculated ED_{01}s based on the southwestern Taiwanese data (Chen et al. 1985, 1992; Wu et al. 1989), the Chilean data (Ferreccio et al. 2000), and the northeastern Taiwanese data (Chiou et al. 2001). It calculated cancer risk estimates for the southwestern Taiwanese data (Chen et al. 1985, 1992; Wu et al. 1989), and the Chilean data (Ferreccio et al. 2000) discussed below. Cancer risks were not estimated for northeastern Taiwan (Chiou et al. 2001) because of instability of the model calculated with the small number of cases in that study.

COMPARISONS OF RESULTS OF DOSE-RESPONSE ASSESSMENTS

Estimates of Effective Dose for a 1% Response: ED_{01}

Doses of an agent associated with the onset of a defined rate of observable response in a study population (often termed ED_{01} when referring to a response rate of 1%) can be useful in several key respects in risk assessment. The ED_{01} can be used as a point of departure for extrapolation to lower doses (typically to the origin) when insufficient data exist to characterize the shape of the dose-response curve in a region. They can also be used to assess a margin of exposure (MOE) between a dose with observed adverse effects and the level of exposure that exists in the general population. A MOE is calculated by dividing the dose associated with a defined level of response, such as a 1% (ED_{01}) or 10% (ED_{10}) response in animal or epidemiological studies, by the actual or projected human exposures (EPA 1996)—deciding which level of response to use (e.g., 1% or 10%) is a policy choice that depends, in part, on the size and quality of the epidemiological or animal data sets available. Therefore, the smaller the MOE is for a given population, the closer the population exposures are to exposures shown to have an adverse effect. The MOE can provide risk managers with information about the extent of apparent protection for the population. The MOE approach is complementary to more traditional approaches for determining a safe level of exposure, each approach providing different information to the risk managers (Presidential/Congressional Commission on Risk Assessment and Risk Management 1997a,b). Because the human epidemiological data set for arsenic encompasses exposure levels close to those for which the subcommittee calculated ED_{01}s, the subcommittee elected to present its ED_{01}s rather than ED_{10}s used in EPA's margin-of-exposure analyses.

Table 5-3 presents the ED_{01}s estimated by the subcommittee (based on mortality or incidence data, depending on the study) for the Chilean data (Ferreccio et al. 2000), the northeastern Taiwanese data (Chiou et al. 2001), and the southwestern Taiwanese data (Chen et al. 1985, 1992; Wu et al. 1989). Those values were estimated using a number of statistical models to fit the data, including additive and multiplicative models using linear or logarithmic terms in dose. The ED_{01}s were estimated using the published or calculated relative risk values and a modification of the BEIR IV (NRC 1988) formula, as described in Chapter 5. Despite the variability, it is evident that most of the

$ED_{01}s$ are less than a factor of 10 higher than the current U.S. maximum contaminant level (MCL) of 50 µg/L.

The subcommittee determined $ED_{01}s$ (i.e., the dose at which there is a 1% response in the study population) for various studies using a number of statistical models. For example, the estimated $ED_{01}s$ from the Chilean study on lung cancer ranged from 5 to 27 µg/L, depending on the exposure data used.

The previous Subcommittee on Arsenic in Drinking Water estimated $ED_{01}s$ of 404 to 450 µg/L, depending on the model used, for arsenic and male bladder cancer mortality. Those values are approximately within the range of $ED_{01}s$ estimated by this subcommittee. However, because the ED_{01} values reported by the current and prior subcommittees were derived using different biostatistical approaches, they are not directly comparable. The ED_{01} values in NRC (1999) reflect a 1% increase relative to background cancer mortality in Taiwan, whereas the current subcommittee's approach, using a modification of the BEIR IV analysis (NRC 1988), reports $ED_{01}s$ based on a 1% increase relative to the background cancer mortality in the United States. This is an important difference because the background rates for lung and bladder cancer are substantially different between Taiwan and the United States. Background rates for lung cancer in the United States are approximately 3- and 2.3-fold higher than in Taiwan for females and males, respectively; and bladder cancer risks are approximately 1.4- and 3-fold higher in females and males, respectively, in the United States when compared to Taiwan.

Cancer Risk Estimates

The subcommittee presents the theoretical lifetime excess cancer risks for lung and bladder cancer incidence for the U.S. population (females and males calculated separately) at fixed arsenic concentrations in drinking water of 3, 5, 10, and 20 µg/L. Table 6-1 presents the maximum-likelihood estimates (MLEs) of the risk of bladder and lung cancer combined based on the data from southwestern Taiwan. Estimates calculated using the U.S. background cancer incidence data or Taiwanese background cancer incidence data are presented. The U.S. background cancer incidence data is taken from SEER (2001). The Taiwanese background cancer incidence data were estimated by multiplying the subcommittee's corresponding U.S. lifetime incidence rate (Tables 5-7 and 5-8) by the ratio of the Taiwanese annualized rate (You et al. 2001) to the U.S. annualized rate (Ferlay et al. 2001). The relatively small confidence limits around the MLE (+/- less than 12% of the MLE) reflect the

TABLE 6-1 Theoretical Maximum-Likelihood Estimates of Excess Lifetime Risk (Incidence per 10,000 people) of Lung Cancer and Bladder Cancer for U.S. Populations Exposed at Various Concentrations of Arsenic in Drinking Water[a]

Arsenic Concentration (μg/ L)	Bladder Cancer				Lung Cancer			
	U.S. Background Rate[b]		Taiwanese Background Rate[c]		U.S. Background Rate[b]		Taiwanese Background Rate[c]	
	Females	Males	Females	Males	Females	Males	Females	Males
3	3.6	6.8	2.3	2.0	5.4	4.0	1.8	1.7
5	6.0	11	3.8	3.2	8.9	6.8	3.0	3.0
10	12	23	7.5	6.8	18	14	6.2	6.1
20	24	45	15	13	36	27	12	11

[a] Estimates were calculated using data from individuals in the arsenic-endemic region of southwestern Taiwan and data from an external comparison group from the overall (mostly unexposed) southwestern Taiwan area. The risks are estimated using what the subcommittee considered reasonable assumptions. (A U.S. resident weighs 70 kg compared with 50 kg for the typical Taiwanese, and the typical Taiwanese drinks just over 2 L of water per day compared with 1 L per day in the United States. Therefore, it assumes that the Taiwanese exposure per kilogram of body weight is 3 times that of the U.S. population.) It is possible to get higher and lower estimates using other assumptions. Risk estimates are rounded to two significant figures. All 95% confidence limits are less than +/– 12% of the maximum-likelihood estimate and are not presented. It should be noted that those confidence limits are a function of sample size and are not indicative of the true uncertainty associated with the risk estimates. The individual risk estimates for bladder and lung cancer were added together to estimate combined risks.
[b] Risks are estimated using the U.S. background cancer rate (SEER 2001).
[c] Risks are estimated using the Taiwanese background cancer rate. The Taiwanese background cancer incidence data were estimated by multiplying the subcommittee's corresponding U.S. lifetime incidence rate (Tables 5-7 and 5-8) by the ratio of the Taiwanese annualized rate (You et al. 2001) to the U.S. annualized rate (Ferlay et al. 2001).

relatively large sample size, and they are not indicative of the true uncertainty associated with the risk assessment discussed in Chapters 4 and 5. The MLEs of the lifetime excess risks for combined lung and bladder cancer incidence for females range from 9 per 10,000 from exposure to drinking water with arsenic at 3 μg/L to 60 per 10,000 from exposure to drinking water with arsenic at 20 μg/L. The corresponding risk estimates for males are 11 to 72 per 10,000. Those values are estimates of the combined lifetime excess risk of lung and bladder cancer (incidence) in a given population following lifetime exposure to arsenic in drinking water at the given concentration.

As presented in Chapter 5, the subcommittee used data from a study performed in northern Chile (Ferreccio et al. 2000) to estimate the theoretical lifetime excess risk of incident lung cancer in U.S. males and females at arsenic concentrations in drinking water of 3, 5, 10, and 20 µg/L. Using the peak period of arsenic exposure in Chile from 1958 to 1970 as a dose metric, the resulting estimates for excess lung cancer incidence in the United States were 3 to 4 times higher than the risks derived from the Taiwanese data. In contrast, when the dose metric used in the Chilean data was the average arsenic concentration in drinking water from 1930 to 1994, the corresponding risk estimates were an order of magnitude higher.

The previous Subcommittee on Arsenic in Drinking Water presented lifetime excess cancer risk estimates for bladder cancer mortality in males based on its analyses of the southwestern Taiwanese data (Chen et al. 1985, 1992; Wu et al. 1989). Some of those risk estimates are presented in Table 5-1. Those risks were estimated using an external comparison population and a multiplicative linear model. At an arsenic concentration of 50 µg/L of drinking water, the excess risk of bladder cancer mortality for males was estimated to be 10 to15 per 10,000 (NRC 1999). Assuming linearity and dividing by 5, that corresponds to a mortality risk estimate of 2 to 3 per 10,000 at 10 µg/L. If the U.S. mortality rate for bladder cancer is 20% (SEER 2001), that corresponds to an estimated risk of bladder cancer incidence of 10-15 per 10,000. Using the southwestern Taiwanese data, this subcommittee's estimate for lifetime excess bladder cancer incidence in males in the United States at an arsenic concentration of 10 µg/L is 23 per 10,000 (see Table 6-1). Therefore, although some analytical approaches were different, the estimates for bladder cancer risk in males for arsenic at 10 µg/L of drinking water determined by the subcommittee in this report are generally consistent with those presented in the previous NRC report.

As discussed in Chapter 5, EPA did not present theoretical lifetime excess cancer risk estimates for arsenic in drinking water in its notices in the *Federal Register* (2000, 2001). The risk estimates it presents (EPA 2001) are adjusted for the occurrence of arsenic in U.S. drinking water; consideration of such an adjustment is beyond the charge to this subcommittee. It is not possible to directly compare the theoretical lifetime cancer risks estimated by this subcommittee with those presented by EPA. The different assumptions used by EPA (2001) and this subcommittee are presented in Table 6-2.

The subcommittee did, however, use a linear extrapolation from the $ED_{01}s$ estimated in the analysis on which EPA based its risk estimates (Morales et al. 2000) to estimate the theoretical lifetime excess bladder and lung cancer risks at 3, 5, 10, and 20 µg/L, presented in Table 5-2. Thus, the subcommittee

TABLE 6-2 Summary of Assumptions Used by EPA and the Subcommittee for Dose-Response Analyses and Their Impact on the EPA's Risk Estimates Relative to the Subcommittee's Risk Estimates[a]

Study Parameter	EPA (2001)	Subcommittee	Impact
Choice of End Point	Lung and bladder cancer	Lung and bladder cancer	No difference
Choice of Study	Southwestern Taiwanese cancer mortality data from Chen et al. (1985, 1988, 1992)	Southwestern Taiwanese cancer mortality data from Chen et al. (1985, 1988, 1992)	No difference
Model Choice	Multiplicative Poisson regression model with linear extrapolation	Additive Poisson regression with linear extrapolation; BEIR IV (NRC 1988)	Decrease
Selection of Comparison Group	No external comparison group used	External comparison group used	Decrease
Adjustments for Water Intake	U.S. population: Monte Carlo analysis of CSFII (EPA 2000) water intakes Taiwan population: water consumption is 3.5 L for males and 2.0 L for females	U.S. population: Mean daily average from CSFII of 1 L/day for males and females Taiwan population: exposures equal to 3 times U.S. default value, i.e., 3 L/day for males and females	Decrease
Adjustments for Dietary Intake of Arsenic	Taiwan: Adjusted lower bound estimates to account for arsenic from cooking water by adding 1 L of water; therefore, total water intake for males was 4.5 L/day and for females was 3.0 L/day. Also to account for intake from food directly, multiplied lower bound estimate by fraction of arsenic consumed per kilogram contributed by drinking water	Taiwan: added a constant concentration of arsenic (30 μg/day) to exposure rates for all individuals in study villages	Decrease
Adjustments for Mortality versus Incidence	Used Taiwanese mortality data for bladder and lung cancers; adjusted upper bound by 1.25 for bladder cancer to reflect mortality, assumed all lung cancer is fatal in Taiwan	Used U.S. background incidence data for bladder and lung cancers from SEER (2001) database	Decrease

[a] More detailed information about these assumptions can be found in Chapter 5.
Abbreviations: CSFII, Continuing Survey of Food Intakes by Individuals; SEER, Surveillance, Epidemiology, and End Results.

compared its risk estimates with those estimates calculated from the published analyses (Morales et al. 2000) on which EPA based its risk estimates (Table 5-2). The subcommittee notes, however, that the estimates in Table 5-2 are not adjusted for water consumption or arsenic in food in the same manner as used by EPA, nor by this subcommittee in its analysis in Chapter 5. (The adjustments used by EPA for food and water consumption would decrease the risk estimates.) However, even without those adjustments, the risk estimates on which EPA based its analyses are lower than this subcommittee's estimates, regardless of whether the U.S. or Taiwanese background cancer rates are used to estimate the risks. Several factors contribute to that difference. Unlike the subcommittee's estimates, EPA's analyses were based on estimates that were calculated without using an external comparison population. The subcommittee also used a different statistical method than EPA to estimate lifetime cancer risks. The subcommittee has presented lifetime excess cancer risk estimates calculated using either the U.S. or the Taiwanese background rates; Morales et al. (2000) estimated the $ED_{01}s$ using Taiwanese background rates. The magnitude of the difference between the estimates can be seen in Table 6-1. In addition, the method the subcommittee used to adjust for arsenic in food and its assumptions regarding water intake in the U.S. and Taiwanese populations were different from those used by EPA in its analyses.

It should be noted that the subcommittee was split on whether using the U.S. background rates was preferable to using the Taiwanese background rates for estimating arsenic risks in the United States. Some members of the subcommittee felt strongly that using U.S. background rates was the preferred approach, while others felt that there was not sufficient justification to select one set of background rates over the other, and that both should be presented. Thus, the results from both approaches are presented in Table 6-1. The subcommittee agreed, however, that if there was a multiplicative interaction between a complex array of risk factors, including smoking, that establish the background rates, then using the U.S. background cancer incidence rates would be preferred over the Taiwanese background rates for estimating arsenic cancer risks in the U.S. population.

PLAUSIBILITY OF CANCER RISK ESTIMATES

Upon completion of an assessment of the potential health effects of an environmental contaminant, it is wise to compare the results of the assessment with a real-world situation—that is, the adverse health effects observed among the people most exposed to the contaminant. The key factors triggering public-

health concern regarding arsenic in drinking water have been the high incidences of different types of cancer in populations exposed to increased concentrations of arsenic in drinking water (greater than 100 µg/L) in Taiwan,
Chile, and Argentina. The cancer with the highest increases in relative risk in
these countries is cancer of the bladder.

It has been suggested that, if the risks of bladder cancer from arsenic in
drinking water were indeed as high as estimated in this report (see Table 6-1),
high cancer rates would have been anticipated in areas of the United States
with increased concentrations of arsenic in groundwater, and these high rates
would have readily attracted public-health attention. Some simple calculations demonstrate how risk estimates for low-level arsenic exposure in this
report might be difficult to detect by observing geographical differences in
cancer incidence or mortality. To illustrate that point, the subcommittee used
its risk estimate of 45 per 10,000 for bladder cancer incidence in U.S. males
(based on the Taiwanese data, U.S. cancer incidence data, and a ratio of 3 for
water ingestion on a per-body-weight basis for the Taiwanese population
compared with the U.S. population) exposed to arsenic at a concentration of
20 µg/L (Table 6-1). The lifetime risk of being diagnosed with bladder cancer
in U.S. males is 3.42% for the period of 1996-1998 (or 342 per 10,000) (SEER
2001). An increased risk of 45 per 10,000 over a background risk of 342 cases
in 10,000 males would be difficult to detect. In terms of bladder cancer mortality, if it is assumed that only about one in five bladder cancer cases in the
United States results in death (the ratio of mortality to incidence is approximately 20% for U.S. males, SEER 2001), a lifetime excess risk for mortality
from bladder cancer in U.S. males is about 9 in 10,000 following lifetime
exposure to arsenic in drinking water at 20 µg/L. The subcommittee further
explored how that risk contributes to overall U.S. mortality for bladder cancer.
Lifetime mortality for bladder cancer in the United States for males is 0.72%
(72 per 10,000) for the period of 1996-1998 (SEER 2001). That increase in
mortality risk of 9 per 10,000 would be difficult to detect against that background rate of 72 per 10,000. Indeed, it would represent only about 13% of
the total risk of bladder cancer mortality. Furthermore, the denominator of the
risk estimate for arsenic assumes that all 10,000 individuals are at risk (e.g.,
all consume arsenic at 20 µg/L of their drinking water for a lifetime). Detection is further complicated by the variability in the actual exposure to arsenic
in drinking water (not considered by this subcommittee), the unknown distribution of other risk factors (especially smoking), and the mobility of the U.S.
population. However, if the risks for arsenic-related bladder cancer were

higher than the estimate used in this example, then bladder cancer incidence and mortality at exposures of 20 μg/L would be proportionately higher and thus might be easier to detect in a population. Because background lung cancer mortality is almost 10-fold greater than bladder cancer, it would be even more difficult to demonstrate an association between low concentrations of arsenic in drinking water and lung cancer risk. Therefore, although the subcommittee's risk estimates are of public-health concern, they are not high enough to be easily detected in U.S. populations by comparing geographical differences in the rates of specific cancers with geographical differences in the concentrations of arsenic in drinking water.

In accordance with its charge, the subcommittee has not conducted an exposure assessment and subsequent risk characterization and risk assessment. The theoretical lifetime excess cancer risks estimated by the subcommittee and presented in this report, however, should be interpreted in a public-health context using an appropriate risk-management framework, such as that proposed by the Presidential/Congressional Commission on Risk Assessment and Risk Management (1997a,b).

SUMMARY AND CONCLUSIONS

- The subcommittee's evaluation and analyses of the data from southwestern Taiwan indicate that the lifetime excess cancer risks in the United States for bladder and lung cancers combined at arsenic concentrations in drinking water between 3 and 20 μg/L (ppb) are estimated to be between 9 and 72 per 10,000 people based on U.S. background cancer incidence data. (The corresponding range based on Taiwanese background cancer incidence data is 4 to 24 per 10,000.) These estimates can be interpreted in light of EPA's stated goals for public-health protection (EPA 1992).
- Depending on the dose metric used in the study, excess risk estimates for cancer in the United States derived from a recent investigation in Chile are either similar to or higher than risk estimates derived from the Taiwanese data.
- Although these risk estimates are high, they would not be detected in U.S. populations by comparing geographical differences in the rates of specific cancers with geographical differences in the concentration of arsenic in drinking water.

REFERENCES

Chen, C.J., Y.C. Chuang, T.M. Lin, and H.Y. Wu. 1985. Malignant neoplasms among residents of a blackfoot disease-endemic area in Taiwan: high-arsenic artesian well water and cancers. Cancer Res. 45(11 Pt 2):5895-5899.

Chen, C.J., M. Wu, S.S. Lee, J.D. Wang, S.H. Cheng, and H.Y. Wu. 1988. Atherogenicity and carcinogenicity of high-arsenic artesian well water. Multiple risk factors and related malignant neoplasms of blackfoot disease. Arteriosclerosis 8(5):452-460.

Chen, C.J., C.W. Chen, M.M. Wu, and T.L. Kuo. 1992. Cancer potential in liver, lung, bladder and kidney due to ingested inorganic arsenic in drinking water. Br. J. Cancer 66(5):888-892.

Chiou, H.Y., S.T. Chiou, Y.H. Hsu, Y.L. Chou, C.H. Tseng, M.L. Wei, and C.J. Chen. 2001. Incidence of transitional cell carcinoma and arsenic in drinking water: a follow-up study of 8,102 residents in an arseniasis-endemic area in Northeastern Taiwan. Am. J. Epidemiol. 153(5):411-418.

EPA (U.S. Environmental Protection Agency). 1992. Drinking water; national primary drinking water regulations–synthetic organic chemicals and inorganic chemicals; national primary drinking water regulations implementation. Fed. Regist. 57(138):31797.

EPA (U.S. Environmental Protection Agency). 1996. Proposed Guidelines for Carcinogen Risk Assessment. Fed. Regist. 61(79):17959-18011.

EPA (U.S. Environmental Protection Agency). 2000. Estimated Per Capita Water Ingestion in the United States: Based on Data Collected by the United States Department of Agriculture's (USDA) 1994-1996 Continuing Survey of Food Intakes by Individuals. EPA-822-00-008. Office of Water, Office of Standards and Technology, U.S. Environmental Protection Agency. April 2000.

EPA (U.S. Environmental Protection Agency). 2001. 40 CFR Parts 9, 141 and 142. National Primary Drinking Water Regulations. Arsenic and Clarifications to Compliance and New Source Contaminants Monitoring. Final Rule. Fed. Regist. 66(14):6975-7066. (January 22).

Ferlay, J., F. Bray, P. Pisani, and D. M. Parkin. 2001. GLOBOCAN 2000: Cancer Incidence, Mortality and Prevalence Worldwide, Version 1.0. IARC CancerBase, No. 5. IARC Press, Lyons, France. [Online]. Available: http://www.dep.iarc.fr/globocan/globocan.htm. [Sept. 20, 2001].

Ferreccio, C., C. Gonzalez, V. Milosavljevic, G. Marshall, A.M. Sancha, and A.H. Smith. 2000. Lung cancer and arsenic concentrations in drinking water in Chile. Epidemiology 11(6):673-679.

Morales, K.H., L. Ryan, T.L. Kuo, M.M. Wu, and C.J. Chen. 2000. Risk of internal cancers from arsenic in drinking water. Environ. Health Perspect. 108(7):655-661.

NRC (National Research Council). 1988. Health Risks of Radon and Other Internally Deposited Alpha-Emitters: BEIR IV. Washington, DC: National Academy Press.

NRC (National Research Council). 1999. Arsenic in Drinking Water. Washington, DC: National Academy Press.

Presidential/Congressional Commission on Risk Assessment and Risk Management. 1997a. Framework for Environmental Health Risk Management. Final Report. Vol.1. Washington, DC: GPO.

Presidential/Congressional Commission on Risk Assessment and Risk Management. 1997b. Risk Assessment and Risk Management in Regulatory Decision-Making. Final Report. Vol.2. Washington, DC: GPO.

SEER (Surveillance, Epidemiology, and End Results). 2001. Surveillance, Epidemiology, and End Results Program Public-Use Data (1973-1998), National Cancer Institute, DCCPS, Surveillance Research Program, Cancer Statistics Branch. [Online]. Available: http://www-seer.ims.nci.nih.gov/Publications/. [August 20, 2001].

Wu, M.M., T.L. Kuo, Y.H. Hwang, and C.J. Chen. 1989. Dose-response relation between arsenic concentration in well water and mortality from cancers and vascular diseases. Am. J. Epidemiol. 130(6):1123-1132.

You, S.L., T.W. Huang, and C.J. Chen. 2001. Cancer registration in Taiwan. Asian Pac. J. Cancer Prev. 2(IACR suppl.):75-78.